AN INSTITUTIONAL BASIS FOR ENVIRONMENTAL STEWARDSHIP

The Structure and Quality of Property Rights

ENVIRONMENT & POLICY

VOLUME 35

The titles published in this series are listed at the end of this volume.

An Institutional Basis for Environmental Stewardship

The Structure and Quality of Property Rights

by

Doris A. Fuchs

Ludwig-Maximilians-University,
Munich, Germany

SPRINGER-SCIENCE+BUSINESS MEDIA, B.V.

A C.I.P. Catalogue record for this book is available from the Library of Congress.

ISBN 978-90-481-6166-9 ISBN 978-94-017-0709-1 (eBook)
DOI 10.1007/978-94-017-0709-1

Printed on acid-free paper

Table of Contents

Acknowledgements **vii**

1 Introduction **1**
1.1 Determinants of environmental quality 2
1.2 Property rights and environmental stewardship 5
1.3 The assurance of property rights 7
1.4 The environmental desirability of government intervention 10
1.5 Where do we go from here? 12

2 Determinants of environmental quality **15**
2.1 The Environmental Kuznets Curve 15
2.2 Institutions, social and cognitive factors 31
2.3 Conclusion 42

3 Property rights and environmental stewardship **43**
3.1 Property rights: The fundamentals 44
3.2 Property regimes 49
3.3 The Tragedy of the Commons 51
3.4 Differentiating types of property rights 62
3.5 So what? 64

4 Assurance of property rights and environmental stewardship **67**
4.1 The effect of the assurance of property rights 69
4.2 Determinants of assurance 75
4.3 Assurance of property rights – a Panacea? 79
4.4 Limits to the positive environmental effects of assurance 81
4.5 Conclusion 86

5 The environmental desirability of government intervention **89**
5.1 Policy as a change in property rights 90
5.2 The environmental desirability of intervention 92
5.3 Implications 105
5.4 Conclusion 108

6 Conclusions **111**
6.1 The current stage: What did we learn? 111
6.2 The next stage: Where do we go from here? 114
6.3 The ideal stage: What really needs to be done 116

Bibliography **121**

Appendix A **133**
A1 The dependent variable: Environmental quality 133
A2 The independent variables 140
A3 The econometric technique 148
A4 The evidence 150
A5 Discussion 168

Appendix B **173**
B1 Culture 173
B2 Income distribution 175
B3 Implications 177

Acknowledgements

A book is rarely the result of efforts by an individual, all by herself, removed from the world (preferably on a mountain top), with her thoughts, her pen and paper - or rather her laptop. It was not the case with respect to this book either. Numerous people have in one way or another influenced its creation. First and foremost, I need to thank my teachers and mentors, Daniel Mazmanian, Jacek Kugler, and Monty Hempel, who with their diversity of scholarly interests and approaches allowed me to develop a crucial breadth in training and ideas, and whose challenges and support provided both inspiration and encouragement. Moreover, I am grateful to the various colleagues who at one stage or another have heard or read parts of the argument presented in this book and provided valuable comments and criticism, especially Ronald Mitchell, Thomas Princen, Elinor Ostrom, Hans Bressers, Michael Kraft, Sheldon Kamieniecki, Pam Zeiser, and Cecil Eubanks, Mark Gasiorowski, and Eugene Wittkopf. Along the way, I have benefited from the support of great colleagues and friends at the Center for Clean Technology and Environmental Policy at the University of Twente, the Geschwister Scholl Institute at the University of Munich, and the Political Science Department at Louisiana State University. Ada Krooshoop did a superb job in putting together the final manuscript and deserves big thanks for that. Most importantly, of course, I would not be where I am now without the support of my friends and family, who have always been there through my good and my bitchy :-) times. Thank you.

Chapter 1

Introduction

What is the role of government in environmental politics and policy? The answer to this question used to be relatively clear. Government was to regulate the environmental performance of state and non-state actors, to set standards, impose charges, and establish more or less stringent criteria of acceptable behavior. With the increasing appearance of the issues of globalization and global governance in the political science literature, however, the focus has turned to the role of non-state actors. Academic research and the popular debate have identified business, non-governmental organizations (NGOs), and civil society as such as increasingly important and potentially powerful actors in the political arena. At the same time, some observers have proclaimed the influence of government to be declining. Others have argued that the role of government primarily is changing rather than declining. Those who adopt the latter perspective postulate that the new role for government in this changing political and socio-economic context is to set the framework for action. In this perspective, government is still of fundamental importance for the functioning of the society, the economy, and political decision-making, but its role is different in that government 'merely' provides the institutional framework facilitating desired outcomes.

This book explores what the governmental provision of an institutional framework could look like in the context of environmental governance. It assumes that government can facilitate the sustainability of resource management by state and non-state actors through the structuring of incentives and constraints. Specifically, the analyses combined in the book inquire into the influence of government on actors' environmental performance through modifications in the structure and quality of property rights. In pursuit of their objectives, the analyses establish links between previously unconnected research areas in environmental politics and resource management and integrate disparate debates on environmental governance and quality. The analyses build on previous research on determinants of environmental quality in general, and the environmental implications of property rights in particular. Furthermore, they draw linkages to the literatures on government capacity and collective action. While speaking to all of the various research areas and debates to some extent, the book, at its core, always pursues the question how government can improve environmental stewardship through modifications in the quality and structure of property rights independent of the level of economic development.

Previous research has demonstrated that the answer to this question is not straight forward. While the tragedy of the commons, for instance, frequently does occur, it is by no means a necessary outcome of joint use and management of natural resources. Neither private property rights nor state-ownership promise general solutions to the overexploitation and degradation of environmental resources. Research on the environmental implications of property rights, thus, has to pursue more sophisticated analyses and ask more differentiated questions than the original academic debates (and to some extent still the present popular debates) suggest.

The analyses in this book pursue such questions at two levels. On the one side, the analyses ask how government can improve environmental stewardship through modifications in the institution of property rights at the macro-level. On the other side, the analyses explore how government can influence environmental stewardship through intervention in the structure of property rights at the micro-level. In both cases, the objective is to identify how governments can structure the socio-economic setting of environmentally relevant decision-making so that the potential for the sustainable management of natural resources increases. Thus, while the book considers research on both environmental quality and environmental stewardship, its primary interest is in the latter. In contrast to environmental quality, the concept of environmental stewardship implies agency. Environmental stewardship is a function of decisions by actors, for which government can provide a supportive setting.

1.1 Determinants of environmental quality

In pursuit of its inquiry into the conditions for successful environmental stewardship, the book starts with a look at previous research on the (cross-national) determinants of environmental quality and stewardship in chapter 2. Given a lack of differentiation between the two debates in earlier research, both literatures include relevant information. In the overview provided in chapter 2, the discussion pays particular attention to assessments of the

'Environmental Kuznets Curve,' i.e. the relationship between economic growth and environmental quality. Research on this relationship caused quite a stir in the early 1990s. Many of our local, regional, and global environmental problems today appear to have been exacerbated if not caused by industrialization and the pattern of development today's developed countries have followed: "Currently, one average person from an industrial country consumes up to fifty times as much material and energy as an average person from a developing country" (Sagasti and Colby, 1993: 97). In consequence, it becomes extremely important whether we believe that future economic growth in developed and especially developing countries will lead to increasing or decreasing environmental problems.

Accordingly, the world listened up when scholars proclaimed that economic growth fosters improvements in environmental quality (Grossman and Krueger, 1991, 1995; World Bank, 1992). These scholars conducted studies that showed that the relationship between economic growth and environmental quality can take a variety of forms, but that an Environmental Kuznets Curve is the one form that applies to the majority of indicators tested. An Environmental Kuznets Curve describes a deterioration of environmental quality with initial increases in per capita income levels, and an improvement in environmental quality once a certain level of per capita income has been reached. Some observers interpreted these results to mean that with sufficient economic growth the world could simply 'grow out' of its environmental problems.

Such pronouncements turned out to be premature, however. By now, numerous studies have explored the implications of economic growth for environmental quality. They find that the Environmental Kuznets Curve is less dominant than previously thought and that other relationships, most importantly uniformly deteriorating or improving environmental quality with rising per capita incomes, are powerful and frequent as well. Furthermore, individual research efforts have shown contradictory findings on the economic growth-environmental quality relationship. Finally, newer studies have emphasized an often neglected acknowledgement of the earlier ones. There is nothing automatic about the relationship between changes in per capita income levels and environmental quality. Increases in per capita incomes even at higher levels do not have to lead to improvements in environmental quality. The Environmental Kuznets Curve may be a function of specific historical political, economic, and social conditions.

Thus, while the earlier studies provided valuable insights into the dynamics between environmental quality and economic growth, they have made answering a different question even more important: how can we improve environmental quality at any given level of development? If we do not want to sit back and wait to grow into or out of a specific environmental problem, and especially if such growth will always lead to simultaneous but contradictory developments with respect to environmental quality, we have to identify tools with which we can improve environmental quality independent of per capita income levels. The question we need to answer is: *What are the political, economic, cultural, and/or social factors that allow us to move the Environmental Kuznets Curve (or any other curve) in order to achieve better environmental quality at any given level of development?*

Political scientists, on their part, have attempted to identify institutional, social, and cognitive determinants of environmental quality, which would allow us to find answers to the above question. An interesting question concerns, for instance, the strengths and weaknesses of democracies in dealing with environmental problems. On the one side, democracies might be more responsive to public demands for environmental quality and more transparent and accountable in political and economic decision-making than non-democratic regimes. On the other side, individual rationality may prevent the necessary sacrifice and the solution of the associated collective action problems in democracies. Likewise, elected officials depending on votes and serving the needs of their constituencies may be unwilling to make the hard choices necessary for environmental protection. Similarly important is the question whether the degree of centralization in a political system affects environmental policy outcomes. Again, arguments for the environmental superiority of either institutional structure can be made.

Scholars also pursue broader institutional approaches when assessing the determinants of environmental quality. As part of the neo-institutional tradition, Ostrom (1990, 1998) applies a rules based approach to studying problems of the commons, in particular common property resource management. Other scholars concentrate on questions of institutional capacity rather than structure. Analyzing local environmental policy capacity, Press (1998) argues that the combination of a community's social capital, political leadership and commitment, administrative and economic resources, and environmental attitudes and behavior form the core determinants of environmental outcomes. While a community's local policy capacity is different from the institutional capacity of government, of course, Press' analysis can provide further information on the origins of institutional capacity. Institutional capacity measures have become very popular in political science in general. Kugler and Arbetman (1997) and Clague, Keefer, Knack and Olson (1995, 1996), for instance, have developed aggregate measures of political capacity that have been applied to a variety of political and economic contexts.

In terms of social determinants of environmental quality, chapter 2 draws attention to the role of income distributions. After all, if per capita Gross Domestic Product (GDP) has such a significant impact on natural resource management, then one needs to look at income distributions to better approximate how much income the majority of people in a country really have. A highly uneven distribution of wealth in a middle income country could mean that small wealthy parts of the population have functioning sewage systems and access to safe drinking water, and are concerned for clean air, clean rivers, and beautiful landscapes in the proximity of their homes, while large, poor parts of the population lack the financial means to improve their environment or are forced to destroy it in the pursuit of survival.

Finally, chapter 2 discusses the role of cognitive factors in determining environmental policy decisions. Cognitive approaches relate differences in environmental policy and/or quality between countries to the cognitive maps, frames, discourses, or cultures of the central actors or societies. The fundamental insight of these approaches is that the behavior of actors rests on their subjective interpretation of reality. In a somewhat similar vein, the advocacy coalition framework emphasizes the role of beliefs in determining policy outcomes.

Chapter 2's overview of the institutional, social, and cognitive determinants of environmental quality does not cover all of the numerous research themes and factors scholars have identified in this field, but only concentrates on a few clusters of the most important ones. Research in this area is ongoing. Furthermore, while quantitative assessments are not the primary objective of this book, the appendices provide the results of preliminary econometric analyses of the environmental impact of the most interesting of these influences on environmental quality (at a given level of development): institutional capacity, democracy and level of centralization, post-materialism, and income distributions.

1.2 Property rights and environmental stewardship

In chapter 3, then, the book turns to its central focus of analysis: the environmental implications of property rights. Laying the ground for the analyses of chapters 4 and 5, chapter 3 provides an overview of the theoretical and empirical findings on the meaning of property rights for environmental stewardship. As a first step, the chapter delineates the fundamentals of a theory of property rights, discussing the nature of property rights, dynamics associated with changes in property rights, and the meaning of property rights for social outcomes in general and harvesting and investment decisions in particular. Following the economic approach to property rights, the book treats property rights as bundles of rights to attributes of a resource (or good in general) which structure actors' choice sets.

The literature on environmental resource management traditionally differentiates between four types of property regimes. The most common categorization of property regimes distinguishes between private property, common property, open-access, and state-ownership. The use of this categorization of property regimes has declined in the literature, however, due to the lack of precision of the categories. Scholars have turned to more differentiated analyses of the form and distribution of specific property rights. Likewise, the present analyses of the environmental implications of property rights will suggest a different approach on the basis of the delineation of two dimensions of property arrangements in chapter 5.

As pointed out above, the tragedy of the commons argument is central in the literature on the environmental implications of property rights and captures the potential for environmental (and economic) unsustainability associated with an insufficient definition of property rights. The tragedy of the commons argument depicts the collective action problems resulting from limited excludability and partial rivalness of resource stocks and flows. The associated collective action problems reflect a lack of individual control over the long-term use and management of natural resources. The fundamental underlying dynamic is to maximize private short-term benefits, as these can be 'secured,' which then leads to the patterns of overuse identified above. The traditional economic argument thus predicts that an absence of (private) property rights will lead to overexploitation of environmental resources and, in consequence, to their degradation or destruction.

The tragedy of the commons argument clearly is not without weaknesses. Scholars have raised numerous criticisms that should cause concern about applying the concept too broadly and too easily. The most fundamental criticism applies to the narrowness of perspective of the analysis. Pivotal determinative factors underlying the structure of the 'game,' such as the impossibility of communication, for instance, are absent in many common property management situations. Furthermore, critics find Hardin's (1968) propositions to be too simplistic, as numerous other factors such as the technology of public supply, the payoff structure, length of the 'game,' the modeling of cooperation and defection as a continuous variable, and institutional rules have an impact on outcomes. Changes in these allow results other than the depletion or degradation of the resource. Finally, the literature on common property resources provides empirical evidence that Hardin's tragedy of the commons generally applies to open-access resources, but often is not the outcome for common property resources. Scholars have identified a range a factors that foster the ability of joint appropriators from a resource to overcome their collective action problems including the homogeneity of the group, the presence of leadership, the ability to exclude outsiders, and most fundamentally the ability to communicate and to learn.

Those criticisms of the tragedy of the commons argument cannot rule out that the tragedy of the commons can occur, however. Similarly, the applicability of some of the game theoretic extensions frequently is limited in the context of environmental problems. The assumption of repeated games is inappropriate, for instance, when we consider most forms of pollution abatement. In addition, the empirical evidence provided by case studies of common property resources concentrates mostly on small scale resources, and its generalizability to large scale resources is debatable due to differences in transaction costs and consequently, payoff structures. On the one side, then, the simple relationship between the exploitation and maintenance of a resource and a lack of private property rights cannot be taken for granted. On the other side, both extensions of the theory of collective action and the empirical evidence on common property resources illustrate that collective action problems and the failure of appropriators to resolve them do exist and are particularly prevalent in cases in which the group of appropriators of open-access resources is too large and heterogeneous.

Finally, some scholars suggest that it is useful to differentiate between different types of property rights. Schlager and Ostrom (1992), for example, argue that it is important to distinguish between owners, appropriators, claimants, and authorized users of a resource as well as between de jure and de facto rights. The strength of such a categorization of types of property rights is that it allows to capture differences in the extent of control provided by property rights and therefore differences in incentives to manage resources efficiently. Critics of this categorization, however, may argue that the differences within each of the categories still are too large. Furthermore, Schlager and Ostrom themselves demonstrate constraints set by this schema when they find that de facto rights of authorized users might lead to a more efficient management of a resource than de jure rights of proprietors.

In sum, theoretical arguments and empirical evidence exist for the potential environmental benefit of clearly delineated property rights as well as the potential of sustainable joint management of resources by numerous appropriators. The common ground in the literature, however, appears to be that open-access resources are the most vulnerable resources with respect to unsustainable management. In addition, traditional categories in property rights research need to be reassessed and transformed to fit the needs of current research questions.

1.3 The assurance of property rights

These findings, then, provide the basis for the analyses in chapters 4 and 5. Those chapters explore the environmental implications of property rights from the two perspectives pointed out above. First, chapter 4 explores how assurance of property rights can influence the potential for environmental stewardship. In this inquiry, the analysis remains at the macro-level of the earlier studies on the roles of factors such as per capita income levels or regulatory systems. In a second step, chapter 5 moves to the micro-level. Following the tradition of Ostrom (1990, 1992, 1998) and others, the chapter examines the implications of different characteristics of property arrangements for the environmental desirability of government intervention with respect to specific natural resources. The results derived in both inquiries hopefully allow us to improve on the unsatisfactory answers from the studies on the relationship between environmental quality and economic growth. They indicate how we can move the Environmental Kuznets Curve and any other curve, and improve environmental quality at any given level of development.

Chapter 4 analyzes the potential and limitations of assurance for improving environmental stewardship. The primary conclusion reached is that the nature of the impact of assurance on environmental stewardship is a function of the relationship between the 'environmental' and 'economic' values of a resource. While having the potential to improve environmental stewardship in many instances, assurance of property rights does not necessarily induce environmentally sustainable outcomes as long as 'environmental' and 'economic' values of resources significantly diverge, i.e. as long as environmental costs and benefits are not fully internalized.

The assurance of property rights influences environmental stewardship primarily through its impact on transaction costs and expectations. By lowering transaction costs, assurance has the potential to increase the extent to which decision-makers define and enforce property rights. In turn, if property rights are increasingly defined, a reduction in open-access resources and externalities results. Assurance of property rights is similarly important for the creation and for the protection of property rights. Owners will choose to protect their rights only as long as the benefits of doing so outweigh the costs. Those resources for which the expected benefits do not balance the costs involved in protecting the rights to the resource will become open-access resources. The situation is even more clear if we think of governments that are not incapable of protecting property rights, but are predatory and unwilling to do so. The higher the chances that resources will be appropriated by the government, the higher the incentives of owners to withdraw benefits without regard for the sustainability of their actions.

In sum, assurance of property rights is positively related to the extent and specificity of property rights' definition and enforcement by individuals. If resources for which no property rights are defined and/or enforced are most vulnerable to the tragedy of the commons, assurance of property rights is of fundamental importance for environmental stewardship. Overall, then, the willingness and capability of government to assure property rights can be paramount for guiding consumption and investment decisions made by appropriators from natural resources in the interest of environmental stewardship.

Assurance of property rights depends on the government's willingness and capacity.[1] The former is a function of government preferences. The capacity of government to assure property rights, in turn, is a function of the overall resources available to government, the efficiency of their utilization, and competing demands for government resources. These resources include material resources from taxes and mineral resources, as well as 'human' resources in form of government support from the population as well as from powerful forces in the state such as the military or economic elites. Highlighting the role of government willingness and capacity, empirical studies have identified as link between periods of political instability and lack of management of state property, assurance of (private) property rights and environmental degradation.

Assurance of property rights is not a panacea for environmental stewardship, however. The property rights and wise use movements in the United States have shown that a high level of assurance of property rights can sometimes hinder the government in pursuing environmental objectives. Courts have ruled against government on the basis that 'taking without compensation' is unconstitutional and required substantial payments of indemnities to property owners for the imposition of use restraints on private property for the benefit of the environment. Thus, assurance of property rights is not a sufficient condition for improvements in environmental stewardship.

Under what conditions does assurance of property rights help environmental stewardship, then? The arguments and examples presented in chapter 4 suggest that assurance of property rights is most desirable and helpful for environmental stewardship with respect to resources whose utilization is dependent on their environmental health. If property owners can achieve substantially higher economic benefits from a resource use that is opposite to its environmental benefits, however, assurance may have a negative impact on environmental stewardship. Thus, the deciding factor that underlies the impact of assurance on environmental stewardship is the difference between the 'environmental' and the 'economic' values of a resource. If economic and environmental values of a resource are close, the decision-maker's maximization of expected utility can imply a maximization of environmental stewardship, and therefore the most efficient property

[1] See chapter 4 for a discussion on the extent to which government can be replaced by other actors in providing assurance.

regime in economic terms may also be the most desirable one environmentally, i.e. lead to the least environmental overexploitation or degradation. If, however, the difference between the two values is large, the maximization of expected utility is likely to result in the maximization of environmental degradation, and therefore the economically most desirable property regime will not be the environmentally most desirable one. While being an effective tool to provide a higher potential for environmental stewardship in many instances, the assurance of property rights, thus, cannot guarantee an environmentally sustainable outcome as long as environmental and economic values of resources significantly diverge. Importantly, the argument is at no time a simple advocacy of private property rights. Rather, the analysis shows that the assurance of property rights can benefit as well as hurt environmental stewardship under all forms of property regimes. In fact, both analyses of the environmental implications of property rights in this book demonstrate that one cannot unequivocally advocate any one property regime for the benefit of the environment.

1.4 The environmental desirability of government intervention

Turning to the environmental implications of property rights at the micro-level, chapter 5 highlights that the environmental desirability of public intervention in property rights is a function of the following factors: state capacity and commitment, collective action problems among the appropriators from the resource, and the relationship between the economic and environmental value of the resource. Building on the literatures on (the security of) property rights, common property regimes, and political capacity, the analysis develops a theoretical 'model' of the relationship between property arrangements and environmental stewardship. The inquiry is motivated by a desire to integrate different strands of research on the environmental implications of property rights that communicate insufficiently with each other. In consequence, scholars fail to take advantage of insights that can be gained from a synthesis of findings. Based on such a synthesis, the 'model' developed in this chapter leads to an innovative analysis of the environmentally desirable level of state-intervention in private ownership of natural resources.

The analysis utilizes a few simple and yet fundamental changes compared to earlier inquiries into the environmental implications of property rights and property regimes. First, the study transforms the traditional categorization of property regimes into private property, common property, and open-access. Instead, it utilizes a continuous variable based on the group of appropriators from the resource and based on the various factors influencing the success of governance among them. Fundamentally, the variable summarizes the level of collective action problems among the group of appropriators.

Secondly, the analysis replaces the traditional fourth property regime of state-ownership with a continuous variable of degree of state-intervention. This step is based on the realization that neither private ownership nor state-ownership of resources is ever complete. Rather than understanding - as has been the tendency - state-ownership as a category of property regimes, it is more useful to see state-ownership as a high degree of state-intervention (as opposed to a low degree of state-intervention) in property ownership. This variable is then laid over the variable capturing the collective action problems among the appropriators so that property arrangements are identified by the characteristics of the appropriators from the resource as well as the degree of state-intervention in the property rights of these appropriators.

Thirdly, having introduced the variable of state-intervention, the study adds the governing bodies' capacity and commitment to sustainability to the analysis. The former is an important determinant of the environmental implications of state-intervention in property arrangements in so far as it identifies a government's ability to protect and enforce its own rights. Government capacity is a necessary condition for a high potential of government intervention for environmental stewardship because only a capable government will be able to enforce rights and regulations and be able to protect or provide an environmental good, which is threatened by private interests. In the absence of government capacity to enforce its own property rights, 'state-owned' property automatically becomes de facto open-access. Government commitment, in the context of this chapter, identifies whether the government is likely to maximize public environmental welfare or private economic benefit. If a government is interested in deriving the maximum economic benefits from environmental resources under its control rather than protecting the public good of environmental quality, government intervention is similar in its environmental implications to unregulated private property arrangements. The capacity and environmental commitment of a government, then, determine the environmental implications of government intervention.

Finally, the analysis takes up the idea from chapter 4 that the difference between the maximum economic value of a natural resource and its economic value when used in an environmentally desirable way is a pivotal determinant of the environmental implications of a given property arrangement. Private property rights, for instance, only raise the potential for environmental stewardship in cases in which the correlation between the economic and environmental values of a resource is high. Indeed, if the maximization of expected utility from the resource leads to a dramatic divergence from the environmentally desirable use, private property, being the allegedly *economically most efficient property regime*, might just lead to the *most efficient environmental degradation.*

These four variables, the level of collective action problems among appropriators, the degree of government intervention, government capacity and commitment to sustainability, and the relationship between the environmental and economic values of a given resource, then, determine the environmental implications of any property arrangement. Since the analysis is primarily interested in potential policy implications, the degree of government intervention is treated as the dependent variable in the discussion. Furthermore, the analysis treats the two situations of a large versus a small divergence between the environmental and the economic values of a resource separately.

Based on the three 'independent variables,' then, the analysis demonstrates how the various arguments and evidence from previous debates and research can be integrated into one cohesive picture. The argument also makes explicit the potential and limitations of various property rights based strategies for improving environmental stewardship. First, the analysis highlights that for government intervention to be environmentally desirable, governments have to be both committed to environmental goals and capable to effectively pursue their policy objectives. Given political conditions around the world, this requirement may place severe limitations on the desirability of government intervention on behalf of the environment. Secondly, the analysis highlights the potential and limitations of privatization for environmental stewardship. In other words, the argument illustrates under which circumstances an increase in individual control over natural resources, as advocated by proponents of privatisation, will make government intervention more or less environmentally desirable. Most fundamentally, the analysis highlights the importance of the relationship between a resource's environmental and economic values as this can dramatically alter the environmental desirability of government intervention in a given situation.

1.5 Where do we go from here?

What do the analyses in this book tell us? What have we learned about government's abilities to influence environmental stewardship through changes in the quality and structure of property rights? What do we still need to find out about these abilities? What further questions have arisen with respect to the institutional basis of environmental governance or approaches to its assessment? The conclusion will attempt to provide answers to these questions.

Chapter 2

Determinants of environmental quality

This chapter provides a brief survey of core determinants of environmental quality identified by previous research. The first part of the presentation focuses on the role of per capita income levels, since the latter received a lot of attention in the 1990s. The analysis shows that the level of economic development as captured by per capita income levels for numerous reasons cannot provide a basis for governance strategies to improve environmental stewardship and quality. The chapter then offers a short discussion of the potential influence of some of the institutional, cultural, and social factors examined as potential determinants of environmental quality by scholars and the frequent lack of agreement on the impact of these factors, before the following chapter turns to the role of property rights as the central focus of this book.

2.1 The Environmental Kuznets Curve

The industrialized countries are responsible for a large share of the degradation of the world's ecosystems. Much of this degradation is directly related to their level of development and the associated production and consumption structures. All of these industrialized countries continue to pursue economic growth as one of their top priorities.[1] In addition, an even bigger number of underdeveloped and developing countries with their associated large share of the world's population hopes to achieve a level of

[1] Economic growth and development are not the same, of course. However, in so far as they are confused in the popular debate and by political and economic decision-makers, this analysis treats them interchangeably.

development comparable to that of the industrialized countries. With such worldwide aspiration for economic growth all around, the pivotal question appears to be, if there is a necessary and identifiable relationship between economic growth and environmental quality.

Accordingly, the world listened up when in the early 1990s scholars announced that the empirical evidence showed that a country's economic development would eventually lead to improved environmental quality (Grossman and Krueger, 1991; World Bank, 1992). These scholars argued that environmental quality would tend to decline with initial growth but would improve after a certain level of per capita incomes was achieved; the relationship between per capita income levels and environmental quality thus taking the form of an inverted U-curve, or an Environmental Kuznets Curve as it has come to be called in the literature. These findings were in stark contrast to earlier concerns about the impact of economic growth on the world's ecosystems raised, for instance, by the Club of Rome. Furthermore, these findings were politically very convenient as they implied that the widely accepted pursuit of economic growth could render unpopular environmental agreements and pressure, on developing countries in particular, unnecessary.

A decade later, some of this euphoria regarding the Environmental Kuznets Curve has died down. Numerous theoretical, methodological, and ecological criticisms of arguments for the existence and political reliance on an Environmental Kuznets Curve have been raised. Yet, the discussion is by no means concluded. Studies continue to explore the influence of per capita income levels on environmental quality, and political and economic decision-makers continue to argue for the positive influence of economic growth on environmental quality in the popular debate. Thus, it makes sense to start this overview of determinants of environmental quality with the research on per capita income levels. The following discussion begins with a more in-depth review of the early studies before it turns to the criticisms of those studies and results of more recent analyses.

In the early 90s, sophisticated empirical studies with data sets covering a large number of countries over several years provided the basis for academic and popular discussions of the influence of rising per capita incomes on the environment (recall, for instance, the debate on the North American Free Trade Agreement, NAFTA). The work by Grossman and Krueger (1995, 1994, 1991), Jänicke, Binder, Bratzel, Carius, Joergens, Kern, and Mönch (1995), and Shafik and Bandyopadhyay (1992) provide the most prominent early examples of such analyses.[2] These authors investigate the determinants

[2] See also de Bruyn 1997, de Bruyn, van den Bergh, and Opschoor 1998, de Bruyn and Opschoor 1997, Dinda, Coondoo, and Pal 2000, Gangadharan and Valenzuela 2000,

of differences in environmental quality among nations for a variety of indicators, and specifically analyze the relationship between environmental quality and income levels. Significantly, all three research endeavors find per capita income to be the most important determinant of environmental quality, explaining in many cases a substantial share of the variance.

According to these studies, the relationship between per capita incomes and environmental quality takes a variety of forms. For many indicators of environmental quality, the findings suggest that the relationship between per capita income and environmental quality follows the form of a Kuznets Curve (graph a). In other words, environmental quality initially worsens with increases in per capita income, but improves with further increases once a certain level of income has been reached. Other indicators, however, provide evidence of a uniform deterioration in environmental quality with rising per capita incomes (graph c), or alternatively a general improvement in environmental quality with rising per capita incomes (graph b).

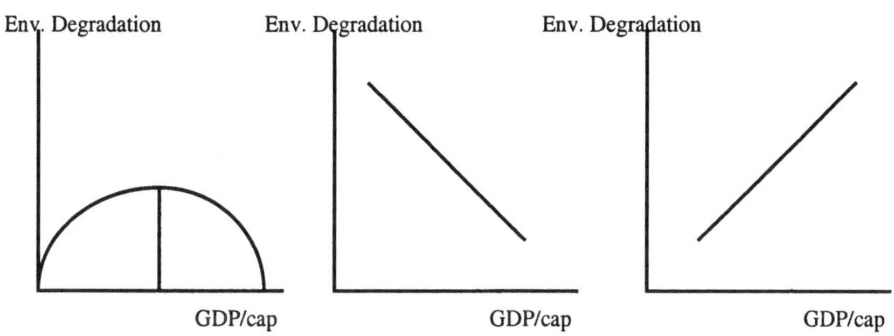

graph a graph b graph c

Grossman and Krueger (1991, 1993, 1994, 1995) first started analyzing the meaning of economic growth for environmental quality in the context of the NAFTA debate, attempting to determine what an increase in growth as a consequence of the free trade agreement would mean for Mexico's environmental quality. They soon extended this analysis to a more general level by including a larger number of indicators (see Table 2.1). In their 1995 *Quarterly Journal of Economics* article, Grossman and Krueger examine the reduced-form relationship between the level of a country's per capita income and environmental quality for three indicators of urban air pollution and ten indicators of water pollution (covering the state of the oxygen regime, pathogenic contamination, and heavy metals).

Kaufman et al. 1998, Moomaw and Unruh 1998, Selden and Song 1994, Stern, Common, and Barbier 1996, Suri and Chapman 1998, Torras and Boyce 1998.

Table 2.1. Results of Grossman and Krueger (1995)

		Environ-mental Indicator	Unit of Measurement	Result	Significant at	Peak (St. Error)
Air pollution		SO_2[3]	$\mu g/m^3$	∩[4]	1%	4,053 (355)
		Heavy Partic.	$\mu g/m^3$	\	1%	NA
		Smoke	$\mu g/m^3$	∩	1%	6,151 (539)
Water pollution	Fecal Contamination	Total Coliform[5]	log(1+Y)[6]	~[7]	<1%	3,043 (309)
		Fecal Coliform	log(1+Y)	-- \	<1%	7,955 (1296)
	State of Oxygen	Dissolved Oxygen	$\mu g/m^3$	∩[8]	<1%	2,703 (5328)
		BOD	$\mu g/m^3$	∩	<10%	7,623 (3307)
		COD	$\mu g/m^3$	∩	22%	7,853 (2235)
		Nitrate	$\mu g/m^3$	∩	<1%	10,524 (500)
	Contamination with Heavy Metals	Lead	$\mu g/m^3$	\[9]	signif.	1,887 (2838)
		Cadmium	$\mu g/m^3$	--[10]	signif.	11,632 (1096)
		Arsenic	$\mu g/m^3$	∩	signif.	4,900 (250)
		Mercury	$\mu g/m^3$	--	./.	5,047 (1315)
		Nickel	$\mu g/m^3$	--	./.	4,113 (3825)

[3] Based on median daily concentration in year.

[4] Emissions appear to increase again at high values of GDP/cap, but results are not reliable because of small number of observations.

[5] Based on mean value of pollution in year.

[6] Where Y = concentration level ($\mu g/m^{3}$).

[7] Possibly spurious relationship.

[8] Results are reported consistent with other indicators, i.e. environmental quality first deteriorates with increases in per capita incomes, and then improves after a certain level of per capita income has been reached. This means that the scale for dissolved oxygen is inverted, as lower levels of dissolved oxygen indicate a lower level of environmental quality, i.e. more pollution. The 'peak' at a per capita income of $ 2,703 therefore is in reality a trough.

[9] Grossman and Krueger describe the relationship as "mostly downward sloping" (1995: 369).

[10] The authors report the curve to be "flat, with perhaps a slight downturn at high levels of income" (ibid.).

sample: varying in size dependent on Global Environmental Monitoring System (GEMS) data availability, including up to 58 countries with up to 47 sites, 1977-88 for air quality data, 1978-89 for water quality data.
source of environmental indicators: GEMS (World Health Organization (WHO) and the United Nations Environmental Programme (UNEP); and the Environmental Protection Agency of the United States (US), as data managing agency).
method: reduced form equations, panel, random effects, Generalized Least Squares (GLS).
income variables: G, G^2, G^3, G', G'^2, G'^3, (where G' = average GDP/cap over past three years), test combined significance of GDP terms.
control variables: time trend, plus (where applicable): location within city, land use nearby, population density in city, dummy for location of city on coast line, dummy for location of city within 100 miles of a desert (for suspended particulate matter (SPM)), dummy for type of measuring device, mean annual water temperature in river (for indicators of water quality).

Consistent with their earlier publications, Grossman and Krueger (1995) find that while per capita income levels have a strong influence on environmental quality, rising per capita incomes do not necessarily worsen environmental quality. On the contrary, for most indicators, including Sulfur Dioxide (SO_2), smoke, Dissolved Oxygen (DO_2), biochemical oxygen demand (BOD), fecal coliform levels (COD), nitrates, and arsenic, environmental quality appears to initially deteriorate with increasing per capita income, but improves after a certain level of per capita income has been reached. Grossman and Krueger find that the turning points vary, but that they are below a per capita income of \$ 8,000 in most cases: “For a country with an income of \$ 10,000, the hypothesis that further growth will be associated with deterioration of environmental conditions can be rejected at the 5 percent level of significance for many of our pollution measures” (1994: 19).

Grossman and Krueger’s analysis presents a thorough and methodologically sophisticated study of the relationship between per capita income and environmental quality. The authors fail, however, to explicitly address the influence of political and social factors on this relationship. While pointing out that there is nothing automatic in the dynamics between per capita income levels and environmental quality, i.e. improvements in environmental quality at higher levels of per capita incomes reflect a conscious political and social choice, Grossman and Krueger fail to identify through which tools this choice might be pursued. This leaves some of the most important questions for policy decisions unanswered.

In an effort similar to Grossman and Krueger’s, Shafik and Bandyopadhyay (1992) study the relationship between economic growth and environmental quality for up to 149 countries during the period 1960-1990 (see Table 2.2). In addition to indicators of air and water quality, they place emphasis on ‘non-pollution’ measures of environmental quality. Their

Table 2.2. Results of Shafik and Bandyopadhyay (1992)

Environmental Indicator	Lack of Drink. Water	Lack of Sanitat-ion	SPM	SO_2	□ in Forest 61-86	Annual Deforestation	Dissolved Oxygen	Fecal Coliform	Municipal Waste	CO
Unit of Measurement	% of pop. w/out access	% of pop. w/out access	micro g/m³	micro g/m³	$\log\left(\frac{(FA_{61}-FA_{86})*100}{FA_{61}}\right)$	log [FA_{n-1} - FA_n]	millig/m³	# p. 100 milliliter	kg/cap	annual metric tons/cap
GDP/cap	\	\	∩	∩	--	--	/[1]	~	/	/
Peak			3,280	3,670						7,000,000
Economic Growth			–					+	+	
Investment				+		+		+		+
Energy Pricing			–	–[2]	–	–				–
Trade I[3]			./.[4]	–		–				
Trade II								–		–
Trade III				+		–				+
Debt				–						+
Political Liberties				+		+	–			+
Civil Liberties				+		+				–

sample: up to 149 countries, 1960 - 1990.

sources of environmental data: World Bank, Bank Economic and Social Database, CCIW and MARC (GEMS data), Marland, 1989, OECD, WRI.

method: panel regressions, controlling for country specific "fixed effects," such as endowment (constant varies)

income variables: log Y, $\log Y^2$, $\log Y^3$, (where Y = GDP/cap in purchasing power parity terms)

control variables: time trend, plus (where applicable) city and site dummies (central residential, central commercial, suburban residential, suburban commercial)

1. Results are reported in a manner consistent with other indicators, i.e. environmental quality worsens with increases in per capita incomes.
2. If a time trend is included, the result is the opposite (+).
3. Trade I is operationalized as the ratio of the sum of imports and exports to GDP. trade II is based on the Dollar Index of trade orientation: a higher index implies a more open trade regime. trade III is based on the Parallel Market Premium, a generalized measure of trade distortions.
4. The results show switching signs in different specifications.

dependent variables include measures of deforestation, access to drinking water or sanitation, as well as waste generation.

Like Grossman and Krueger, Shafik and Bandyopadhyay find an environmental Kuznets curve for a number of indicators. The broad range of indicators used, however, also allows Shafik and Bandyopadhyay to identify other forms of the relationship between environmental quality and per capita income, which highlights that skepticism is appropriate for simple generalizations. Their statistics exhibit, for instance, that access to clean water and sanitation tends to uniformly improve with rising incomes, while other indicators such as dissolved oxygen in rivers, carbon emissions, or municipal waste appear to worsen at higher income levels. Furthermore, the results do not show a significant relationship between per capita income levels and deforestation.

Besides per capita income, Shafik and Bandyopadhyay estimate the influence of a range of economic variables such as economic growth, rates of investment, energy pricing, trade policy, and debt burden on environmental quality. Their results suggest a variety of forms of relationships between environmental quality and these variables, some counter-intuitive. In most cases, however, the economic variables turn out to be insignificant. Unfortunately, Shafik and Bandyopadhyay limit their analysis of political indicators to an assessment of the impact of Gastil's index of political and civil liberties on environmental quality, which they find to be insignificant for most environmental indicators except dissolved oxygen, SO_2, and deforestation.

Finally, while pursuing a somewhat different question, Jänicke et al. (1995) perform an analysis quite similar to that of Grossman and Krueger and Shafik and Bandyopadhyay. Jänicke et al. set out to determine conditions for success in environmental policy by analyzing differences in performance among 32 industrialized (in some samples 24 OECD) countries. Yet, the results of their research speak to the same agenda. After providing a thorough literature review of empirical comparative analyses of environmental policy in general as well as political, cultural, and social factors identified as determinants of environmental quality, the authors conduct several regression analyses for a range of environmental indicators, including Suspended Particulate Matter (SPM), Nitrogen Oxides (NO_x), Volatile Organic Compounds (VOC), wastewater treatment, waste generation, and land use. They concentrate, then, on explaining the differences in performance in SO_2 emissions (see Tables 2.3 and 2.4). In the latter case, they assess a much wider range of potential determinants than either of the two research teams discussed above, noting, however, that for many of the political conditions of successful environmental policy discussed in the literature, such as the ability to develop a consensus,

Table 2.3. Results of Jänicke et al. (1995) I

Indicator	Unit of Measurement	Result
SO_2	annual kg/cap	∩ [11]
SPM	annual kg/cap	\ [12]
NOx	annual kg/cap	/
CO	annual kg/cap	∩
VOC	annual kg/cap	/ [13]
Energy Consumption	annual kg TOE/cap	∩ [14]
Water Treatment[15]	% of population served	\
Total Water Abstraction	annual m^3 /cap	∩
Municipal Waste	annual kg/cap	/
Major Prot. Areas	% of territory	\

sample: 32 industrialized countries (24 OECD countries)
sources of environmental data: OECD Environmental Data Compendium, OECD National Accounts, Statistical Abstract of the United States
method: correlations

decentralization, or openness of the decision-making institutions, hardly any adequate indicators exist.

Like the other two research teams, Jänicke et al. find per capita GDP to be correlated with all of the eight indicators of environmental quality, with the relationships taking the forms of a 'Wohlstandswende' (an Environmental Kuznets Curve), a 'Wohlstandsentlastung' (a positive relationship between environmental quality and per capita income levels) or a 'Wohlstandsverschmutzung' (the negative equivalent). In terms of the influence of other factors on environmental quality (as measured by the SO_2 emissions coefficient), Jänicke et al. argue that the influence of environmental policy is mostly overshadowed by other developments, for instance in energy conservation or transportation growth, and therefore cannot be quantitatively measured.[16]

[11] Jänicke et al. point out that the result appears to be a decrease in SO_2 emissions with increasing per capita incomes at first glance, because most of the countries in the sample reached the peak in SO_2 emissions before 1970.

[12] The authors emphasize that this result applies to stationary sources, as emissions from non-stationary sources continue to increase with rising per capita incomes.

[13] As some of the wealthier countries have started to reduce VOC emissions, the relationship might follow the form of an inverted U-curve.

[14] While no absolute reductions can be recognized, energy consumption rose only marginally in the wealthier countries in the sample.

[15] Results reported in manner consistent with presentation of the other results: environmental quality improves with increases in GDP/cap, i.e. the percentage of the population served by water treatment plants increases as per capita incomes rise.

[16] Moreover, Jänicke et al. emphasize the limited reliability of quantitative analyses in the environmental field, because of the current paucity of cross-national data. The authors ask the reader to understand the postulated relationships as models for interpretation or

Table 2.4. Results of Jänicke et al. (1995) II: SO_2 [17]

	GDP/ cap	Unemploy-ment	Inflation	Structural Ec. Conditions	Public Concern [18]		Income Distribution	Research & Development
fossil SO_2[19] emissions coeff.	\	+	–	–	– (1982)	./. (1992)	+	+ [20]
R-squared	40.4	47	28.2	19.6[21]	15.5	2.8	0[22]	37.8
adjusted SO_2 coefficient[23]	\	+	–	–	./.	./.	+	+
R-squared	21.6	14.1	32.8	9.4	0	2.2	0	34.6

Factors not found to be correlated with SO_2 emissions:
Δ in GDP 1970-1990; Investment; SO_2 in 1970; Population Density; Urbanization; Post-Materialism; Country Size (Area, Population); Openness (Trade); Power of Communities (decentr.); Parties in Government; Institutionalization (Date)

Jänicke et al.'s major contribution to this area of inquiry derives from their consideration of a wide range of determinants of environmental quality, including social, political, cultural, and demographic factors. Such a broad effort, however, necessarily also entails certain weaknesses. Apart from

descriptive statements but not as laws that empirical tests failed to reject (Jänicke et al., 219).

17 *Sources*: Organization for Economic Cooperation and Development (OECD), Environmental Data Compendium 1993, the International Energy Agency (IEA), Energy Balances of OECD Countries / Energy Balances of non-OECD Countries, 1993. SO_2 emissions are calculated from industrial statistics rather than measured. Jänicke et al. report their correlation results for SO_2 for various samples: all countries, OECD (always without Australia, Greece, and New Zealand), plus OECD without countries with extreme patterns such as Spain, or Turkey. I present only their results for "all countries" here. The high degree of variance in results depending on which countries were included in the sample, however, indicates a limited robustness of these results (see results for Trade Balance and Structural Conditions of Economy).

18 Results presented here refer to concern about the local environment. Concern about the national environment showed similar results.

19 The indicator is calculated as the ratio of SO_2 emissions from fossil fuels (- industrial emissions) to fossil fuel consumption. However, it is likely to overestimate influence of variables tested. The limitations of the indicator are: difference in sulfur content in different fossil fuels, impact of share of transportation caused emissions in total emissions, importance of climatic conditions (little need for heating, large potential for hydro-power).

20 As Jänicke et al. point out, the relationship might simply be a function of the higher research intensity in wealthier countries, as R&D expenditures fail to explain any variation in SO_2 emissions between the 16 wealthiest countries.

21 The R squared rises to 35%, if only OECD countries are analyzed.

22 The lack of significance is likely a function of the limited variance in the sample.

23 The indicator is based on the residual from regressing the share of coal in fossil fuel use on the fossil SO_2 coefficient, i.e. deviation of the SO_2 coefficient from the value which could have been based on the coal share in fossil fuel consumption (likely to underestimate influence of variables in contrast to the first indicator).

questions of whether the variables used actually capture what they were intended to capture, Jänicke et al.'s analysis is weakened by the lack of a consistent and parsimonious theoretical model.[24]

There is little disagreement among these studies on the importance of per capita income levels as determinants of environmental quality. Yet, they leave important questions unanswered and reveal new ones. Criticism has centered on three issues: the lack of consistent evidence on the existence of an Environmental Kuznets Curve, the absence of a comprehensive explanation of the relationship between per capita income levels and environmental quality, and the missing insights on how governments can actively attempt to improve environmental quality at a given level of per capita income.

To what extent does the empirical evidence provide support for the existence of an Environmental Kuznets Curve and a consistent influence of per capita income levels on environmental quality in general? Little agreement on the empirical evidence exists apart from the point that per capita income has an important influence on environmental quality (see also Ekins, 1997; Fuchs, 1996[25]; Stern et al., 1996). For several indicators of environmental quality, the three studies focused on here arrive at different results regarding the form of this relationship (see Tables 2.5, 2.6, and2.7 7). Shafik and Bandyopadhyay, for instance, find that the relationship between CO_2 emissions and per capita income levels is uniformly positive while Jänicke et al. postulate it to follow a Kuznets Curve. Similarly, Grossman and Krueger find a Kuznets Curve for DO_2, while Shafik and Bandyopadhyay find oxygen levels to uniformly worsen with increases in per capita income. Likewise, Shafik and Bandyopadhyay argue that SPM reflects the pattern of an Environmental Kuznets Curve, while Jänicke et al. identify a positive relationship between environmental quality in terms of SPM and per capita income levels. Grossman and Krueger, at the same time, differentiate between heavy particulates and smoke, and find both a Kuznets Curve for smoke and a positive relationship for heavy particulates. Furthermore, these

[24] Based on their findings for the different variables, the authors eventually develop a LISREL model. In this model, the authors identify "Wohlstand" (welfare) as a latent variable which through its effect on employment and inflation determines SO_2 emissions. The other variables are postulated not to have a direct influence on SO_2 emissions. It is not obvious, however, what the theoretical basis for this claim is or why inflation and unemployment should be functions rather than determinants of "Wohlstand." Most importantly, the authors claim of a direct relationship particularly between unemployment, inflation and SO_2 emissions is surprising, given that Jänicke et al. themselves point out that this relationship might be spurious and based on the generally superior economic performance of capable governments (1995: 154).

[25] See also Appendix A.

Table 2.5. Summary of Results I: Similar Indicators

Environment. Indicator	Study	Unit of Measurement	GDP/ cap	Peak	Countries (cities, river stations) in Sample	Years	Method	Data Sources	Additional Variables Found to Be Significant[1]
	G&K	μg/m³	\∩[2]	6,151[3]	29 (36) for heavy part., 19 (18) for smoke	1977-1988	panel, random effects, GLS	GEMS (EPA)	
SPM	S&B	micro g/m³	∩	3,280	31 (48)	1972-1988	panel, fixed effects	GEMS (MARC)	Ec. Growth (-), Energy Pricing (-)
	J. et al.	annual kg/cap	\		32 industrialized (24 OECD) countries	1970, 1990	correlations	OECD, USA[4]	
CO	S&B	metric tons/cap	/	7 Million	118-153	1960-1989	panel, fixed effects	Marland 1989	Investment (+), Energy Pricing (-), Trade II (-), Trade III (+), Debt (+), Pol. (+) & Civil Lib. (-)
	J. et al.	kg/cap	∩		32 industrialized (24 OECD) countries	1970, 1990	correlations	OECD, USA	
	G&K	μg/m³	∩		< 58 (<287)	1979-1990	panel, random effects, GLS	GEMS (CCIW)	
Dissolved Oxygen[5]	S&B	millig/m³	/		27 (57)	1979-1988	panel, fixed effects	GEMS (CCIW)	Political and Social Liberties (-)
	G&K	log(1+Y)	-- \[6]	7,955	42	1979-1990	panel, random effects, GLS	GEMS (CCIW)	
Fecal Coliform	S&B	# per 100 milli-liter	~		25 (52)	1979-1988	panel, fixed effects	GEMS (CCIW)	Investment (+), Trade II (-)
	S&B	kg/cap	/		39	1985	panel, fixed effects	OECD, WRI	Economic Growth (+)
Municipal Waste	J. et al.	kg/cap	/		32 industrialized (24 OECD) countries	1975, 1990	correlations	OECD, USA	

1. Variables having a negative influence on the environmental indicator in question are marked by (-), variables having a positive influence by (+).
2. The results are "\" for heavy particulates, and "∩" for smoke.
3. Peak applies to smoke.
4. USA = Statistical Abstract of the United States.
5. Results are reported in a manner consistent with other indicators, i.e. according to the findings of Grossman and Krueger environmental quality first deterorates with increases in per capita incomes, and then improves after a certain level of per capita income has been reached, while Shafik and Bandyopadhyay find environmental quality uniformly decreasing with rising per capita incomes. This means that the scale for dissolved oxygen is inverted, as lower levels of dissolved oxygen indicate a lower level of environmental quality, i.e. more pollution.
6. GDP/cap is associated with constant levels of fecal coliform up to per capita incomes of $ 8,000, and falls sharply thereafter.

divergences in results translate to the policy variables. Jänicke et al., for example, find that investment and trade 'openness' have no impact on SO_2 emissions, while Shafik and Bandyopadhyay claim a significant positive relationship for the former and a significant negative relationship for the latter.[26]

These differences in findings might result from a range of factors. A comparison between the results for SPM, for instance, reveals that the differences could originate in a convoluted environmental indicator. Similarly, the differences perhaps are caused by the sample compositions. Jänicke et al., for instance, might be finding an improvement of environmental quality with rising per capita incomes for SPM rather than a Kuznets Curve because the sample includes only countries with high levels of per capita income that passed the 'peak' before the sample period. In sum, the differences in results not only highlight the importance for further research, but also remind us of the need for caution when analyzing data and interpreting findings.

In addition, these comparisons pinpoint the need for the development of a theoretically driven model. Shafik and Bandyopadhyay's findings of the significance of different policy variables for different environmental indicators paint a colorful but scientifically unstructured picture. Similarly, the lack of empirical support, according to Shafik and Bandyopadhyay's and Jänicke et al.'s analyses, for the impact of most of the policy variables parallel the frequent lack of a compelling theoretical foundation for the form and combination in which they are introduced in regressions and correlations. In several cases, the authors themselves suggest that even the 'significant' relationships might be spurious. Likewise, scholars argue that the empirical results may be biased by missing variables, such as the composition of production and consumption, international trade, power and income inequalities, the density of economic activity, rates of economic growth, energy prices, and external shocks (Rothman and de Bruyn, 1998). Fundamentally, the reduced form models used for instance by Grossman and Krueger and Shafik and Bandyopadhyay provide very little information on how the turn in pollution is actually achieved. The Kuznets Curves found, for instance, may record only a displacement of dirty industries.

[26] De Bruyn and Opschoor (1997), by the way, find an n-shaped relationship for different indicators of environmental pressure, suggesting the possibility of a 'relinking' of economic growth and environmental degradation at high income levels.

Table 2 6. Summary of Results II for SO_2

Study	Unit of Measurement	Countries (Cities) in Sample	Years	Data Sources	GDP/cap	Peak
Jänicke et al.	annual kg/cap	32 industrialized, 24 OECD[1]	1970, 1990	OECD, USA[2]	$\cap$	
S & B	microg/m^3	31 (47)	1972-1988	GEMS (MARC)	$\cap$	3,280
G & K	µg/m^3	42 (47)	1977-1988	GEMS (EPA)	$\cap$ [3]	4,053

Study	Invest-ment	Unemploy-ment	Inflation	Debt	Trade I, Openness	Trade III, Market Prem.	R&D	Income Distrib.	Industrial Structure	Energy Pricing[4]	Pol. & Civil Liberties	Public Concern
Jänicke et al.	./.	+	+		./.		+ [5]	+ [6]	–			– ./. [7]
S & B	+			–	–	+				– [8]	+	

Factors found not to be correlated with SO_2 emissions:

Economic Growth (Δ in GDP 1970-1990)*
Trade Balance**
SO_2 in 1970**
Population Density**
Urbanization**
Post-Materialism**
Country Size (area, population)**
Decentralization**
Parties in Government**
Institutionalization (date)**
Trade II (Dollar Index)***

* = by both Shafik and Bandyopadhyay, as well as Jänicke et al.
** = by Jänicke et al.
*** = by Shafik and Bandyopadhyay

1. For Australia, New Zealand, and Greece, no reliable data were available.
2. USA = Statistical Abstract of the United States.
3. Grossman and Krueger point out that emissions appear to increase again at high levels of per capita incomes. Because of the limited number of observations in this range, however, the results are not reliable.
4. "Energy Pricing" is based on electricity tariffs.
5. As Jänicke et al. point out, the relationship might just be a function of a generally higher research intensity of wealthier countries.
6. The relationship between income distributions and SO_2 emissions is not statistically significant in this analysis, however, because of the limited variance in income distribution between the countries in the sample.
7. Public concern about local environmental quality was found to have an effect in 1982 but not in 1992. Jänicke et al. identified a similar pattern for concern about national environmental quality.
8. If a time trend is included, the result is the opposite (+).

Table 2.7. Summary of Results III: Additional Indicators

	Environmental Indicator	Study	Unit of Measurement	GDP/cap	Peak	Countries (Cities, River Stations) in Sample	Years	Method	Data Sources
	NOx	J. et al.	annual kg/cap	/		32 industrialized (24 OECD) countries	1970, 1990	correlations	OECD, USA
Air Pollution	Smoke	G&K	μg/m3	∩	6,151 (539)	19 (18)	1977-1988	panel, random effects, GLS	GEMS (EPA)
	VOC	J. et al.	annual kg/cap	/[1]		32 industrialized (24 OECD) countries	1980, 1990	correlations	OECD, USA
	(total) Water Abstraction	J. et al.	annual m^3/cap	∩		32 industrialized (24 OECD) countries	1970, 1990	correlations	OECD, USA
	Water Treatment	J. et al.	% of populat. served	\		32 industrialized (24 OECD) countries	1970, 1990	correlations	OECD, USA
Water Consumption and Pollution	BOD[3]	G&K	μg/m	∩	7,623 (3,307)		1979-1990	panel, random effects, GLS	GEMS (CCIW)
	COD	G&K	μg/m^3	∩	7,853 (2,235)		1979-1990	panel, random effects, GLS	GEMS (CCIW)
	Nitrates	G&K.	μg/m^3	∩	10,524 (500)		1979-1990	panel, random effects, GLS	GEMS (CCIW)
	Coliform (total)	G&K	log(1+Y)	~[2]	3,043 (309)	22	1979-1990	panel, random effects, GLS	GEMS (CCIW)
	Arsenic	G&K	μg/m^3	∩	4,900 (250)		1979-1990	panel, random effects, GLS	GEMS (CCIW)
	Cadmium	G&K	μg/m^3	--[3]	11,632 (1,096)		1979-1990	panel, random effects, GLS	GEMS (CCIW)
	Lead	G&K	μg/m^3	\	1,887 (2,838)		1979-1990	panel, random effects, GLS	GEMS (CCIW)
	Mercury	G&K	μg/m^3	--	5,047 (1,315)		1979-1990	panel, random effects, GLS	GEMS (CCIW)
	Nickel	G&K	μg/m^3	--	4,113 (3,825)		1979-1990	panel, random effects, GLS	GEMS (CCIW)
Deforestation	Deforestation (annual)[4]	S&B	$\log[FA_{n-1} - FA_n]$[5]	--		66	1962-1986	panel, fixed effects	BESD
	Deforestation (Δ 61-86)[6]	S&B		--		77	1961-1986	panel, fixed effects	BESD
Conservation	Major Protected Areas	J. et al.	% of territory	\		32 industrialized (24 OECD) countries	1970, 1990	correlations	OECD, USA
Energy	Energy Consumption	J. et al.	annual kg TOE/cap	∩		32 industrialized (24 OECD) countries	1970, 1990	correlations	OECD, USA
	Lack of Safe Drinking Water	S&B	% of population	\		43-44	1975, 1985	panel, fixed effects	BESD
	Lack of Sanitation	S&B	without access	\		55 – 70	1980, 1985	panel, fixed effects	BESD

1. Possibly ∩.
2. Possibly spurious relationship.
3. "Flat with perhaps a slight downward turn at high levels of income" (1995: 369)
4. Additional variables found to be significant: Investment (+), Energy Pricing (-), Trade I and III (-), Debt (+), Political and Social Liberties (+).
5. Where FA = forest area in thousands of hectare, $n \in [1968, 1986]$.
6. Additional variables found to be significant: Energy Pricing (-).

Progress in this area of inquiry will also depend on further enhancing the compatibility and comparability of studies by finding a common ground on questions of measurement and methodology. Critics of these early studies have found their choices on functional forms of models and the focus on emissions rather than stock levels debatable (De Bruyn and Opschoor, 1997; Rothman and de Bruyn, 1998). De Bruyn, van den Bergh, and Opschoor (1998), for instance, argue that the estimation of a more dynamic model shows that economic growth has a direct positive (i.e. growth) effect on the level of emissions. While scholars might not always be able to agree on what type and form of environmental indicators are most appropriate, such as city-based or national data, per capita emissions or ambient levels (and data availability further limits our options) or on the functional form of the model, we have to create awareness for such differences and their consequences. Furthermore, we have to develop a consensus on how to interpret results derived from the utilization of specific indicators. Sulfur dioxide emissions, for instance, are predominantly from industrial sources, and changes in emissions might therefore be primarily a function of structural changes in the economy, as more polluting industries move abroad and are replaced by the service sector. Likewise, Jänicke et al. rely on estimates of SO_2 emissions that are calculated based on industrial structure and therefore do not reflect actual emissions - this implies difficulties in interpretation of the results.

In addition to the differences in empirical results between the studies, critics highlight the lack of ecological validity and policy relevance of their findings. Scholars have criticized that the studies fail to link pollution patterns to ecosystems resilience and carrying capacities, or more generally to environmental sustainability (Arrow et al., 1995). Likewise, the analyses' tendency to use indicators of single pollutants and materials rather aggregate ones fails to provide a complete picture of the relationship between economic growth and environmental quality, as one cannot exclude the possibility of 'transmaterialization' (de Bruyn and Opschoor, 1997).[27] Finally, the analyses frequently identify the turning point of the Environmental Kuznets Curve at relatively high income levels which do not seem achievable for the majority of the world population (Stern, Common, and Barbier, 1996).

Finally, the studies fail to provide a comprehensive explanation for the variety of forms the relationship between per capita income levels and different indicators of environmental quality takes. Grossman and Krueger 'suspect' that the Environmental Kuznets Curve is a function of increased

[27] These challenges are difficult to overcome at the moment, however. De Bruyn and Opschoor (1997) themselves acknowledge that presently virtually every aggregation would be arbitrary from an environmental point of view.

demands for environmental quality at higher income levels (1995: 17). Shafik and Bandyopadhyay explain the differences in the relationship between environmental quality and per capita income by reference to the effect of environmental quality on human welfare: "Where environmental quality directly affects human welfare, higher incomes tend to be associated with less degradation. But where the costs of environmental damage can be externalized, economic growth results in a steady deterioration of environmental quality" (ibid.: 4). Finally, Jänicke et al. argue that the likely causal pattern behind these divergent paths is the structure of the environmental problem and the ease of solution. They postulate that environmental quality will improve with rising GDP if a problem can be solved through technological add-ons or modified regulation such as the protection and conservation of specific areas. If, however, such a technological or administrative solution is not available or requires significant intervention in markets and established routines, environmental quality will worsen as GDP increases.

A complete explanation, of course, has to include all three aspects: *Given that at higher per capita incomes more resources (financial and technological) are available for environmental protection, and concern about other aspects such as food or shelter is less prevalent,*[28] *improvements in environmental quality will be most likely if the pollution or environmental degradation has significant local impacts and if the problem can be solved through (available) technological add-ons or modified regulation without significant intervention in markets and established routines.*

Both, a look at the theoretical link between per capita income levels and environmental quality, and a critical look at the empirical evidence provided by scientific assessments underscore, then, that Environmental Kuznets Curves exist only for some pollutants, which are mainly those that have local health effects and can be dealt with without great expense (Rothman and de Bruyn, 1998). In addition, a focus on the role of per capita incomes fails to provide information on the opportunity for government to improve environmental quality at a given level of income and thus has to present an incomplete picture of the meaning of economic growth for environmental quality. Important facets of the relationship between the two factors remain undisclosed. Even if per capita income levels have a significant influence on environmental quality, they are certainly not determinative. Countries at similar income levels exhibit different patterns of environmental quality. Thus, scholars have explored the role of other factors, specifically institutional, cultural, and social ones, in affecting environmental quality.

[28] This part of the argument, of course, is in congruence with Maslow's hierarchy of needs.

2.2 Institutions, social, and cognitive factors

Political scientists have explored a wide range of factors as potential determinants of environmental quality. In comparative case studies, in particular, scholars have tried to elucidate the role of institutional, cognitive, cultural, or social characteristics. The following presentation aims to provide a brief overview of some of the interesting research questions and efforts focusing on cross-national determinants of environmental quality. It cannot do justice to any individual research effort nor any research theme, and, due to the necessary brevity of description, takes considerable liberty in the clustering of factors and approaches. Moreover, the presentation delineates the results of empirical assessments of selected institutional, cultural, and social determinants of environmental quality by the author, the details of which are provided in Appendices A and B.

Institutions

In explaining cross-national differences in environmental quality, political scientists have concentrated particularly on the role of institutions. Among those that have received the most attention are regime types and levels of centralization. The debate on the influence of regime types on environmental policy addresses one central question: Are democracies better or worse protectors of the environment (Buell and DeLuca, 1996; Fuchs, 1996; Midlarsky, 1998; Payne, 1995; Press, 1994; Shiva, 1999; Williams and Matheny, 1995)? One side in the debate claims that democracies are incapable of solving the major environmental problems we are facing today, as these problems arise from the liberty awarded to self-interested individuals to pursue their personal gain without concern for the social benefit. Individual rationality prevents the necessary sacrifice and the solution of the associated collective action problems (Hardin, 1968; Heilbroner, 1975; Ophuls, 1977). Furthermore, with elected officials depending on votes and serving the needs of their constituencies, government is unwilling to make the hard choices necessary for environmental protection: "(...) I do not see how one can avoid the conclusion that the required transformation will be likely to exceed the capabilities of representative democracy." (Heilbroner, 1980: 106).

The other (and - especially after the demise of the Soviet Union and reports of environmental conditions there - stronger) side, in contrast, claims that democracies are superior in solving environmental problems because of the higher responsiveness to public demands for a clean environment as well as a higher degree of long-term legitimacy awarded to environmental policies (Press, 1994; Passmore, 1974; Payne, 1995). Both responsiveness and legitimacy are identified as functions of the higher accountability of elected officials, the opportunity to elect a 'green party' into office, the transparency of the political process, and the access of scientific knowledge and technological innovation to the political arena. Furthermore, superior information flows are supposed to produce a more knowledgeable public, which therefore voices superior political demands.

A debate closely associated with the above dichotomy focuses on the environmental implications of the level of centralization of the political decision-making process of a country. In fact, the debates are so closely associated that some authors fail to distinguish between advocating democracy or decentralization, or authoritarian regimes and centralized decision-making respectively. *Centralists* argue that centralized decision-making is associated with a higher potential to achieve the optimal social outcome because of superior information gathering and distribution mechanisms, less coordination difficulties, and the absence of competition between the various units (Heilbroner, 1975; Ophuls, 1977; McIntyre and Thornton, 1978). Ringquist's (1993) finding on the greater importance of federal policies compared to state policies for environmental quality in the 50 US states appears to provide empirical evidence for the centralists' claims. However, as Ringquist himself points out, the large difference in resources between the federal and state programs does not allow such a conclusion.

In contrast, *decentralists* claim that decision-making at lower levels and by more units is better for the environment because of the capabilities of a decentralized system to adjust to local needs and provide for a higher level of participation (Press, 1994; Passmore, 1974; Schumacher, 1989; Goldsmith, 1972; Lovins 1977). Authors working in this field further highlight the benefits of pluralism, and the openness and flexibility provided by federalism (Knöpfel and Weidner, 1985; Margerum, 1995; Nelkin and Pollack, 1981; Scheberle, 1997; Weber, 1998; Ziegler, 1980). In an effort to assess these theoretical claims empirically, Knöpfel and Weidner (1985) find that the absence of specific emission standards and consequent flexibility in pollution abatement produced superior environmental results in France, the UK, and the Netherlands, compared to a high density of centralized federal programs in Germany and Italy. Likewise, Margerum finds that "federalism played a beneficial role by allowing states to develop policies which best fit local economic conditions" when studying subnational support for high-tech industries in the US and Germany (1995: 1).

Related to the decentralization argument, a number of authors in earlier studies attributed a significant influence on environmental policy outcomes to the openness of the political system and the access awarded to certain actors (Enloe, 1975; Kitschelt, 1986; Nelkin and Pollack, 1981; Ringquist, 1993; Solesbury, 1976). Enloe (1975) identifies the access of experts to the political arena as an important condition for successful environmental policies. Likewise, Nelkin and Pollack (1981) point out that a relatively more open political system allowed non-governmental actors as members of the anti-nuclear movement a higher degree of influence in Germany than in France, and therefore led to a different governmental stance on nuclear energy. Similarly, Ringquist (1993) identifies the potential for influence of polluting industries awarded by the political structure in the state as an important determinant of environmental policies in the 50 US states.

As the latter studies highlight, however, arguments linking the openness of a political system to policy outcomes are fundamentally limited. In order to link openness to environmental costs or benefits, we have to make assumptions about the interests of those actors benefiting from the openness of the system. Openness as such does not determine whether a policy outcome is better or worse for the environment, as that will depend not on how many parties have access, but rather which parties have (privileged) access and how the resources between the parties having access are distributed. Such an argument is supported by findings that the strength and character of participating actors, such as the environmental movement and the existence of organized interest groups as well as the potential for mobilization of social groups and opposition parties, have an important impact on environmental policy outcomes (Ringquist, 1993; Reich, 1984).

Finally, comparative studies of determinants of environmental policy have focused on a host of specific characteristics of the institutional structure of a political system, such as the balance of power between governmental actors, the distribution of authority in government, legislative characteristics, political rights, or the political style, for instance, (Böhmer-Christiansen, 1992; Böhmer-Christiansen and Skea, 1991; Dasgupta, Mody, Roy, and Wheeler, 1995; Reich, 1984; Ringquist, 1993; Solesbury, 1976). Böhmer-Christiansen (1992), for example, identifies subtle differences in policy-making in the form of institutional and political attitudes towards science, engineering, and government as causes for the difference between a technology-led proactive acid-rain policy in West Germany and a science-led reactive one in the United Kingdom (UK). Ringquist (1993), in turn, finds legislative characteristics such as the length of the term or the endowment of legislature and authorities to have an impact on environmental policy outcomes. Likewise, Brickman, Jasanoff and Ilgen (1985) contrast a slow, consensus-oriented and careful process of decision making regarding the regulation of chemicals in Europe with a highly-fragmented one based on strict formulization of policies and the pivotal role of competition between interests in the United States.

In broader assessments of the determinants of environmental policy than those provided by most environmental policy studies, Dasgupta et al. (1995)

and Jänicke et al. (1995, see above) empirically test a range of political variables. Dasgupta et al. find freedom of property and freedom of information to have a significant influence on environmental policy, while freedom of print and broadcast media, freedom of peaceful assembly, alternative per capita GDP estimates compiled by the United Nations International Comparisons Program, and measures of contract enforceability and bureaucratic delay from the Business Environmental Risk Intelligence data (BERI) fail to exhibit a statistically significant influence. Jänicke et al. test the impact of decentralization, parties in government, public concern, and the institutionalization of environmental policy on SO_2 emissions, but find only public concern in 1982 to be significant.

In addition, scholars have addressed an adversary versus a consensus-oriented policy style as a determinant of policy outcomes, pointing out the difference between cooperative business-government relationships and conflictual ones in Japan, Germany, France, the UK, and the US (Bardaracco, 1985), or contrasting a conflictual policy style based on strict enforcement in the US with a voluntary compliance with environmental standards in the UK (Vogel, 1993).

Combining various arguments and findings, finally, Jänicke and Mönch (1988) argue that the important political determinant of environmental policy outcomes is the ability of a political system to react to problems. According to them, this ability depends on the openness of the system, which determines its ability to innovate, the political capacity of the system in terms of allowing the pursuit of goals in a coordinated manner over long periods of time, which determines its ability to strategize, and the cooperative patterns of interaction, which determine its ability to integrate diverging interests. As this argument by Jänicke and Mönch most clearly illustrates, however, such potential political determinants of environmental policy necessarily involve problems with empirical specifications. Many of these characteristics of political systems cannot easily be assessed at an aggregate level or be employed in empirical studies that look at more than just a few cases. Even in cases in which scholars have empirically analyzed the impact of these variables on policy outcomes, the results have been ambiguous, and in a number of studies the authors find that significant differences in political characteristics between countries fail to lead to significantly different end results: "At this point, the overall result seems very much like a dead heat. Neither of the two countries seems to have been remarkably more successful than the other in relieving its citizens from the blight of air pollution." (Lundquist, 1980: 194, comparing the influence of differences in policy style between the US and Sweden). While such findings could be the results of difficulties in measuring the respective political factors, the explanation might also be that these political factors by themselves do little to determine environmental policy outcomes.

A much broader understanding of institutions and institutional analysis is provided by Ostrom. The 'institutional rational choice approach' is essentially a 'rule-based' approach, according to which collections of rules together with the characteristics of the actors and the physical conditions of resource settings determine (environmental) governance processes and outcomes (Ostrom, 1990, 1999).[29] Ostrom identifies seven types of rules the combination of which characterizes an "action arena": 'entry and exit' rules (which decide the identity of the actors); 'position' rules; 'scope' rules (which identify the field to which this position relates); 'authority' rules (which delineate the competences of the actors); 'aggregation' rules (which regulate the extent to which permission from/agreement with others is required); 'information' rules (which delineate the availability and dissemination of knowledge); and 'payoff' rules (which describe the costs and benefits of a given action for the various actors). In addition, Ostrom highlights that rule making is nested within rules made in other arenas by delineating different levels of rules ranging from constitutional rules, to collective choice and operational rules. References to Ostrom's work will appear again and again throughout this book due to the close relation to this analysis in terms of logic and focus.

Finally, scholars have focused on the role of institutional capacity rather than institutional structure in explaining policy outcomes. In a different but related inquiry, Press (1998) emphasizes the impact of questions of local environmental policy capacity (rather than general institutional capacity) on local environmental outcomes. He measures capacity in terms of a community's ability to solve collective action problems and improve the sustainability of resource management. If one wants to assess institutional capacity at a more aggregate level, the capacity measures suggested by Kugler and Arbetman (1997) and Clague, Keefer, Knack, and Olson (1995, 1996) provide interesting opportunities. While these measures have not been utilized much in an environmental context, their wide applicability promises useful insights. Kugler and Arbetman define government capacity as a combination of penetration and extraction, both measured as deviations from expected economic performance. Clague et al.'s measure, Contract Intensive Money (CIM) is based on the idea that individuals in environments with insecure property rights will choose to engage only in self-enforcing contracts, which provide less potential for the society to realize gains from

[29] Somewhat related to Ostrom's institutional rational choice approach is Scharpf's (1997) "actor centered institutionalism" which explains outcomes (specifically cooperation) on the basis of actor constellations, the distribution of preferences, and the availability of information.

trade and build a foundation for economic growth. Based on this argument, Clague et al. measure contract-intensive money as the ratio of non-currency money to the total money supply $[M_2\text{-}C]/M_2$, where M_2 is a broad definition of the money supply and C is currency held outside banks.

Social Factors

An additional group of factors which scholars have studied as potential determinants of environmental quality are social factors. These include questions of education, gender, religion, and income distribution, for instance. Education is generally seen as being positively related to environmental values, which then can translate into pressure on the environmental performance of economic and political actors. Since gender and religion are not in the general frame of government intervention they will not be discussed further here.

Income distribution, whose influence on environmental quality has received the least attention but raises particularly interesting questions, requires a few comments. In their 1995 study, Jänicke et al. find skewedness in income distribution to be positively related to SO_2 emissions. Intuitively, one also can explain why a country's income distribution should have a significant influence on environmental quality. If per capita GDP has such a significant impact on how we use our natural resources, then we need to look at income distributions to better approximate how much income the majority of people in a country really have. After all, one of the major flaws of per capita GDP as a measure of welfare is that it glosses over disparities in income and wealth within a country. It does not indicate whether the wealth of a nation is in the hands of a small elite or more evenly distributed among the population.

In the case of environmental quality, a highly uneven distribution of wealth in a middle income country could mean that the majority of the population deals with natural resources at the level of poverty, for instance, as subsistence farmers pursuing slash and burn agriculture. In the extreme case, we can imagine well constructed developments of the small wealthy parts of the population with functioning sewage systems and access to safe drinking water, and a concern for clean air, clean rivers, and beautiful landscapes in the proximity of their homes, while large, poor parts of the population lack the financial means to improve their environment or are forced to destroy it in the pursuit of survival. In terms of natural resources, such uneven distribution of wealth might well mean that large amounts of resources are owned and potentially well-kept by the elite, while large numbers of people do not own natural resources, nor have the means to enforce property rights if there are any.[30] For those people, then, natural resources exist only in the form of open access resources.

Cognitive Factors

Recently, a substantial share of research in political science in general and environmental politics in particular has started to utilize cognitive approaches (Axelrod, 1976; Dryzek, 1997; Fischer, 1995; Fischer and Forrester, 1993; Milbrath, 1993; Schön and Rein, 1994; Thompson, Ellis, and Wildavsky, 1990). According to such approaches, differences in environmental policy and/or quality between countries can be linked to the cognitive maps, frames, discourses, or cultures of the central actors or societies. The fundamental insight of these approaches is that the behavior of actors rests on their subjective interpretation of reality. This interpretation, in turn, is a function of the actors' frameworks of interpretation which provide meaning to observations.[31] Furthermore, these approaches often fundamentally challenge the positivism of the analyses focusing on institutional and social factors discussed above. Differences between these cognitive approaches rests on the extent to which they see frameworks of interpretation as specific to a certain actor or to groups or societies.

A related framework that has become increasingly popular for scholars trying to explain cross-national differences in environmental policy is the advocacy coalition framework (Sabatier, 1988, 1999; Sabatier and Jenkins-Smith, 1991). Similar to the cognitive approaches discussed above, the advocacy coalition framework emphasizes the role of beliefs in determining policy outcomes.[32] According to Sabatier, political convictions or policy beliefs are central to (environmental) policy development because they allow advocacy coalitions to develop around them, which in turn will have a significant influence on policy strategies and outcomes. In this perspective, an 'advocacy coalition' is made up of governmental and non-governmental actors who hold common beliefs and, to a considerable extent, coordinate their activities.[33] Different levels of beliefs exist ranging from the 'deep core' (fundamental values) to the policy core (perception of problems and questions of policy design and process) and 'derived aspects' (elaborations

[30] Consequently, concern for the environment might be limited by the boundaries of development. If the industry owned by the elite pollutes the environment outside of the elite's residential areas, any political response is likely to be small or absent, since the people suffering under the pollution do not have the means and political clout to affect change.

[31] To some extent, these frameworks of observation provide a means to deal with uncertainty.

[32] Sabatier (1999) himself emphasizes the difference between his approach and the cognitive ones mentioned above, which he finds too vague (p. 11).

[33] Recently, Fenger and Klok (1998) have extended Sabatier's model by factoring in the resource dependence of the coalitions.

for the respective situation). Various studies have found that differences and the potential for change in these beliefs can explain differences in environmental policy outcomes.

Finally, the postmaterialism debate addresses the role of values and beliefs in determining environmental quality as well. The debate derives its main impetus from the work by Inglehart (1990, 1971; see also Inglehart and Abramson, 1994). Inglehart argues that changes in the culture of Western, industrialized societies, which were caused by developments in the economic, political and technological environments, have led to corresponding changes in the kind of economic growth these societies are pursuing. Inglehart emphasizes an increased concern for quality of life aspects, including environmental quality, and a lessened focus on economic growth rates: "These cultural changes, in turn are now redirecting the trajectory of advanced industrial society, leading to the de-emphasis of economic growth as the dominant goal of society and the decline of economic criteria as the implicit standard of rational behavior" (Inglehart 1990: 3). Some empirical research finds a strong relationship between post-material values and environmental values (Dalton, 1994; Müller-Rommel, 1990). Jänicke et al. (1995), however, fail to find a statistically significant influence of post-materialism on SO_2 emissions in their cross-national study.

Furthermore, other scholars link higher incomes, education and younger generations to an increasing turn towards environmental values, which then might not be culturally specific but rather depend on social conditions. In addition, Inglehart's claim that the post-materialism in the Western industrialized societies is very different from non-Western societies has been rejected by empirical assessments of environmental attitudes in several non-Western societies, predominantly African ones, which exhibited a level of environmental concern similar to that found in "post-materialist" societies (Gallup, 1989). In sum, while cultural characteristics necessarily interact with the environment in multiple ways, the question is whether Inglehart's post-materialism really captures the gist of this cultural influence.

Apart from post-materialism, authors frequently identify other characteristics as cultural determinants of policy outcomes, although they often fail to differentiate sufficiently between culture, ideology, and ideas. Enloe (1975), for instance, argues that whether environmental problems will be understood and publicly discussed is a cultural characteristic, which in turn will influence environmental outcomes. Likewise, Böhmer-Christiansen and Skea (1991) claim that German history has led to greater pessimism in the culture and therefore a more careful stance towards environmental problems. The problems with such culturally based arguments are well-known. How do we define culture? Which countries or regions belong to the German culture, for instance, and therefore have absorbed this 'pessimism'?

Most fundamentally, the question is what is 'culture' and how far can we trace its impact?[34]

Empirical Evidence

Based on a selection of the institutional, social, and cultural factors discussed above, a range of empirical analyses on cross-national determinants of environmental quality have been conducted (Fuchs, 1996, for a detailed presentation see Appendices A and B). The particular focus of the analyses was on the influence of institutional capacity on environmental quality, but they also considered the implications of regime types, levels of centralization, governmental environmental effort, income distribution, and post-materialist values. The analyses tested the influence of these factors for twelve indicators of environmental quality ranging from air and water pollution to access to clean water and sanitation. The core concern of the inquiry lay in determining whether the identified factors could improve environmental quality at a given level of development, i.e. when controlling for the influence of per capita GDP levels.

In order to assess the influence of institutional capacity on environmental quality at a given level of development, the analysis utilized the measure of institutional capacity developed by Clague et al. (1995) (see Appendix A). This measure, "Contract Intensive Money" (CIM), is based on the argument that governmental institutional capacity will be reflected in the characteristics of contracts in which the economic sector in a country engages.[35] Thus, Clague et al. expect societies with higher institutional capacity to be more willing to rely on enforcement intensive contracts across space and time than societies with low institutional capacity. The latter will tend to rely on self-enforcing contracts, i.e. direct, short-term transactions and barter.

The results of the empirical analysis of the influence of institutional capacity on environmental quality were very encouraging. The analysis found a positive relationship between institutional capacity and the respective indicators of environmental quality for eight of the eleven indicators tested. The results showed institutional capacity to be positively related to access to rural and urban sanitation and safe drinking water, as well as oxygen levels in rivers, and negatively related to most indicators of environmental pollution tested such as smoke, sulfur dioxide, and biochemical oxygen demand in rivers. Furthermore, in most of the

[34] While Elkins and Simeon (1979) address only political culture as one aspect of culture, they highlight these difficulties in defining and analyzing the impact of political culture.

[35] Clague et al. originally intended CIM to measure the security of property rights in a society. For a discussion of whether CIM does not capture a much broader notion of institutional capacity, see Appendix A.

regressions, the measure of institutional capacity performed better than the GDP terms.

These findings, then, supported the hypothesis that improvements in institutional capacity at any given level of development will be associated with improvements in environmental quality. They thereby suggested a way in which governments could actively pursue environmental improvements rather than having to wait for per capita GDP levels to rise. Questions of institutional capacity thus should provide a promising avenue for future research.

With respect to the other factors tested, unfortunately, the results of the empirical analyses were less promising. Reflecting the controversies about the influence of democracy and levels of centralization on environmental quality described above, the empirical assessments of the nature of these relationships was inconclusive (see Appendix A). Utilizing the democracy measure from Gurr's Polity III data set and indicators of centralization from Gastil's Civil and Political Liberties Index, the analyses found a positive influence of democracy and centralization on some indicators of environmental quality and a negative one on others. Most importantly, in the majority of cases, the democracy and centralization measures were not statistically significant.[36] These results suggested that the nature of the relationships needed to be further specified in empirical assessments. Under what conditions are democratic institutions and procedures likely to be able to foster superior environmental quality? Under what conditions is political decentralization more likely to be associated with environmental improvements?

Besides these institutional factors discussed in the literature, the inquiry also aimed to determine whether it is possible to empirically identify a relationship between governmental environmental effort and environmental quality. Based on the coding of national environmental reports prepared for the United Nations Conference on Environment and Development in 1992, Dasgupta et al. (1995) had developed a set of comparative indices for status of environmental policy and performance in air, water, land, and living resources, which the analysis utilized. Interestingly, the analysis could not show that the environmental policy effort is positively related to environmental quality. On the contrary, the results appeared to suggest that poor environmental quality might lead to higher environmental effort. These findings suggested that the relationship between governmental environmental effort and environmental quality might primarily work in the opposite direction, with effort being high when environmental quality is particularly poor and government intervention is necessary. Since the

[36] With respect to democracy, these findings stand in contrast to later studies (Midlarsky, 1998).

empirical analysis in this case was based on cross-sections rather than time series data, unfortunately, it could not assess the temporal dynamics between the two factors.

Finally, the analysis included preliminary empirical assessments on the influence of social and cultural factors, specifically income distributions and post-materialism, on environmental quality at a given level of development (see Appendix B). Given the lack of sufficient data for these two factors, sample sizes in both cases were quite small, so that the results have to be treated more carefully. However, they did provide some insight into the potential relationships. With respect to the influence of income distribution on environmental quality, the analysis found ambiguous results. The findings indicated a positive relationship between environmental quality and income distributions in some cases and a negative one in others. To the extent that a general trend could be identified, the majority of cases suggested that environmental quality is poorer where income distributions are highly skewed. However, the coefficient on the income distribution variable was significant only for one third of the indicators of environmental quality tested, and for those indicators, the relationship between income distributions and environmental quality is positive, i.e. greater divergence in income distributions leads to better environmental quality. For post-materialism, however, the empirical assessment arrived at promising results. Post-materialist values in society were found to be positively related to environmental quality for five of the six indicators of environmental quality tested, in four of those cases the coefficients were statistically significant.

Overall, the only institutional, social, and cultural factors which the empirical analyses showed to be determinative of environmental quality were institutional capacity and post-materialism. These findings were particularly interesting with respect to the relationship between institutional capacity and environmental quality, for which we lacked empirical research. The findings on the influence of post-materialist values on environmental quality reflected the positive results of earlier assessments (Dalton, 1994; Müller-Rommel, 1990). Neither democracy and centralization, nor environmental effort and income distributions could be shown to be determinative factors of environmental quality. Given that cultural values are difficult to influence in the short term, the results suggested improvements in institutional capacity as the most promising avenue for governments to foster environmental quality at any given level of development. Furthermore, these findings promised that further inquiries into the implications of the structure and quality of institutions for the environment would be appropriate and led to the conceptual analyses presented in the next chapters.

2.3 Conclusion

Scholars have identified a wide range of determinants of environmental quality including economic, institutional, cognitive, and social factors. Unfortunately, only institutional capacity appears to provide a sufficiently promising avenue for governance strategies in pursuit of superior environmental conditions. While post-materialist values also can be shown to be positively associated with environmental quality, influencing cultural values in the short-term is difficult if not impossible. The evidence on the environmental implications of democracy is contradictory, and proof of a strong and consistent influence of other factors such as decentralization, governmental environmental effort, and income distributions is lacking. In spite of the euphoria raised by earlier research, the focus on the influence of per capita income levels on environmental quality has turned out to be inadequate for policy relevant research as well. Even if per capita income levels can be shown to be strongly related to environmental quality, the form of the relationship is controversial, appears to be multi-directional, and does not allow us identify opportunities for governments to improve environmental quality at a given level of per capita incomes. Therefore, we need strategies for fostering environmental quality independent of the level of economic development, such as the one suggested by institutional capacity. Further inquiries into environmental implications of factors related to institutional capacity, then, appear the most promising avenue for research on environmental governance strategies. The next chapters pursue such analyses by turning to more specific institutional factors as a basis for governance strategies in pursuit of environmental stewardship: the structure and quality of property rights. Chapter 3 provides an overview of the discussion on the implications of property rights for environmental stewardship.

Chapter 3

Property rights and environmental stewardship

The debate on the environmental implications of property rights has a long tradition. Gordon (1954) and Hardin (1968) are famous for making a case for the positive impact of the definition and enforcement of property rights on environmental stewardship in the early days of the debate. They argue that rational decision-makers will only manage a resource sustainably and make investment and consumption decisions accordingly, if they are reasonably certain that they will be the principal beneficiaries of the pursuit of sustainability. In the logic of this argument, overexploitation and depletion of natural resources are likely in the absence of long-term control, i.e. in the absence of well-defined and well-enforced property rights.

In contrast, many scholars and activists in the environmental community have viewed property rights as a tool allowing property owners to pursue private economic gain at the expense of the public good of environmental quality. They feel that the increase in the owner's control and the reduction in social control often lead to socially undesirable outcomes. As examples, these critics point out cases in which private development in environmentally valuable areas leads to a deterioration of environmental stewardship.

Both sides have a point, a fact which highlights that the relationship between environmental stewardship and property rights is multi-faceted and complex. The implications of a specific property rights arrangement for the environment depend on political, social and economic contexts. Arguments about the environmental implications of property rights need to be much more specific than the original debates suggested. Before introducing such arguments in chapters 4 and 5, however, it is necessary to recapitulate the background for a property rights based approach to environmental stewardship.

To that end, this chapter lays out the fundamentals of a theory of property rights, delineating the nature of property rights, dynamics associated with changes in property rights, and the meaning of property rights for social outcomes in general and harvesting and investment decisions in particular. Furthermore, the chapter introduces the concept of the 'tragedy of the commons' and discusses its implications and weaknesses. In addition, the chapter delineates different options for categorizing property rights and regimes suggested in the literature and identifies the strengths and weaknesses of these proposed categories as well. It is important to emphasize that the approach to property rights predominantly utilized in this chapter and throughout this book is an economic and not a legal or philosophical one.

3.1 Property rights: the fundamentals

Property rights delineate rights of ownership in an asset, which generally include the rights to use and consume the asset, to exclude others from the use of the asset, to change its form and substance, to obtain income from it, and to transfer these rights either in their entirety through sale or partially/temporarily, for instance through rental (Barzel, 1989; Furubotn and Pejovich, 1975; Kasper and Streit, 1998). While property rights may be exclusive, they are generally not unrestricted.[1] Governments, for example, often impose regulations limiting the owners' options in terms of how they can use their resource (a point we will take up again in chapter 6).

Correspondingly, economists argue that property rights should be conceived of as bundles of rights (Barzel, 1989; Kasper and Streit, 1998). With respect to environmental resources, for instance, property rights frequently differ for the *stock* of a resource and the produced *yield* or the goods and services derived from a resource. 'Ownership of the resource' would thus pertain to a specific bundle of rights the owner holds with respect to the resource. The owner may, for example, hold the right to farm the land, but not to kill rare species on the land. The bundle of property rights associated with an asset, ceteris paribus, determines its value in an exchange and consequently the terms of trade (Furubotn and Pejovich, 1975).

[1] This fact is important to remember in the context of debates on the environmentally desirable property regime. In this debate, private property rights are often treated as absolutes, which in reality they rarely are. Rather than having to choose between private property regimes, common property regimes, and state ownership, the imposition of some constraints on private property often is a reasonable alternative.

We usually do not think about the specific bundle of rights we are purchasing when buying a good, because we have a common general understanding of what those rights are. Likewise, in legal terms, the purchase of a good is generally understood to mean the purchase of a given set of rights. Obtaining title to real estate, for instance, generally means that we have the right to live on our real estate, but generally not to kill somebody on our real estate. This societal and legal understanding of which rights are included in our purchase does not negate the economic perspective of the (changeability of) bundles of rights, however. After all, contractual specifications of bundles of rights that differ from the 'normal' case are possible in property transfers, and courts frequently have to deal with cases questioning whether an owner of a good or asset had the right to carry out a particular action. There are limits to the bundling and unbundling of property rights from a legal perspective, of course, as some contractual agreements on specific bundles of rights might be viewed as unconstitutional or immoral.[2] Furthermore, property owners may not be able to unbundle a particular right, often a negative right, i.e. responsibility associated with their property ownership. In general, however, the economic perspective on property rights as bundles of rights cannot be easily rejected.

Notions about which specific rights generally go with the ownership of a good or asset and the extent to which it is possible to unbundle specific rights differ across time and culture. Thus, it is much more common to differentiate between specific rights to attributes of a good or asset in the Common Law tradition than in German law, for instance. This is somewhat ironic, since the Common Law tradition happens to be based on old Germanic legal frameworks (in which this unbundling was possible). In contrast, German law today is essentially based on the Roman concept of property rights, to which the countries on the continent reverted in the context of the 'creation' of the concept of territorial sovereignty associated with the Westphalian peace in the 17th century. Under Roman law, unbundling was generally much more limited, especially with respect to real estate. Thus, the owner of real estate owned everything below the area of land as well as the space above the area of land. In consequence, owning apartments, for instance, was not possible. Despite this difference in traditions, notions of bundling and unbundling are not written in stone, of course. Thus, unbundling is becoming much more common on the European continent these days, while in the United States property owners are fighting against the right of government to interfere with specific rights associated with their property.

2 Such limits on unbundling can also exist in terms of the feasibility of the separation of property rights to different attributes to the resource.

With respect to natural resources, the focus on 'bundles of rights' highlights that property rights to multiple attributes of a resource can exist and be held by different individuals or groups. Even different 'property regimes' (see below) are likely to exist with respect to the attributes of many environmental resources. Property rights and regimes for such a resource thus tend to form a complex structure with several layers and dimensions.

Property rights play a pivotal role in society as they structure the relations between decision-makers with respect to any natural resource.

> For property rights are defined not as relations between men and things, but rather as the behavioural relations among men that arise from the existence of things and pertain to their use. The prevailing system of property rights assignments in the community is, in effect, the set of economic and social relations defining the position of interacting individuals with respect to the utilization of scare resources (Pejovich, 1975: 40).

Property rights, thus, are fundamental in any social context and with respect to any good or asset. As Stubblebine (1975) argues, the definition of property rights becomes necessary as soon as two individuals share a living environment, e.g. with Friday's arrival on Crusoe's island: "Some set of property rights must, and will, be created to condition the relationship between these two individuals - whether that set be characterized as capitalistic, socialistic, or something else" (ibid.: 13). This definition of rights will take place, whether or not the institution of 'property rights' has been intentionally created by the community or by an outside authority and enforcer.

In combination with social norms, available technologies, and resource conditions, property rights yield collective outcomes in social settings (Young, 1994). They determine the allocation and use of resources, composition of output, and distribution of income which result from interacting decision-makers: "Such institutions critically affect decision making regarding resource use and, hence, affect economic behavior and performance. By allocating decision-making authority, they also determine who are the economic actors in a system and define the distribution of wealth in a society" (Libecap, 1993: 1). Certainly, I am not arguing that property rights are the only determinants of the social outcome of multiple interactions between individuals. The alternatives decision-makers face are not only limited by legal constraints such as constitutions, laws and contracts, but also by informal constraints on behavior, such as social rules, taboos, and norms of conduct. Yet, in an increasingly complex world with interactions between decision-makers across space and time, the definition of relationships through property rights provide a foundation for environmental and economic activity.

The institution of property rights can take a variety of forms. Property institutions "range from formal arrangements, including constitutional provisions, status, and judicial rulings, to informal conventions and customs regarding the allocations and use of property" (ibid.). In traditional societies with dense social networks and accordingly low transaction costs, not much in the sense of formal contracts or written laws is required to guide and structure the interaction between individuals. In contrast, the interactions in today's Western societies require elaborate institutional structures in the form of well-specified and well-enforced property rights to reduce transactions costs: "The establishment of such a set of property rights will then allow individuals in highly complex interdependent situations to be able to have confidence in their dealings with individuals of whom they have no personal knowledge and with whom they have no reciprocal and ongoing exchange relationships" (North, 1989: 1320f).

Two major perspectives on changes in property rights and their long-term stability exist. In the economic perspective, property rights are continuously in flux. In this view, property rights are continuously being created, altered, and abandoned in a functioning society (Barzel, 1989). As individuals desire to adjust to changed economic, political, and social conditions, they create demand for a change in property rights until they are satisfied and have reached a new 'equilibrium' position in rights over resources.[3] Property rights to resources, thus, are determined by the interaction of supply and demand in dynamic sequences. This logic is most clearly expressed in Stubblebine's "Axiom of Modification": "Every individual seeks those property rights modifications which he believes will improve his welfare" (Stubblebine, 1975: 15). As utility maximizers, in the economic perspective, people delineate and exercise property rights up to the point where the marginal costs of doing so exceed the marginal benefits, leaving the remainder in the public domain.

> People acquire, maintain, and relinquish rights as a matter of choice (...) People choose to exercise rights when they believe the gains from such actions will exceed their costs. Conversely, people fail to exercise rights when the gains from owning properties are deemed insufficient, thus placing (or leaving) such properties in the public domain. What is found in the public domain, therefore, is what people have chosen not to claim. As conditions change, however, something that has been considered not

[3] Property rights, thus, are continuously determined at the margins in an ongoing endogenous process. (Obviously, property rights are determined exogenously as well, for instance, through political changes such as revolutions or transitions from communist to capitalist regimes or even through 'simple' policy changes.)

worthwhile to own may be newly perceived as worthwhile; conversely, what was at first owned may be placed in the public domain (Barzel, 1989: 65).

Individuals' 'calculations' of the net-benefit of changes in property rights, in turn, are affected by a number of factors such as changes in relative factor scarcities and relative prices, technological changes (affecting the costs of monitoring and enforcing property rights among others), changes in knowledge (changes in production functions, market values), or the opening of new markets, and the interaction between these factors. North, for instance, argues that institutional change in the past has been predominantly caused by changes in relative prices, which in turn are a function of population growth, technological change, and changes in the cost of information, with population growth being historically the single and most important factor (North, 1989: 1324). Feeny, likewise, claims that an appreciation in the relative price of a factor leads to an "increase in the demand for an institution to define property rights in that factor" (Feeny, 1988:273). He draws empirical support for his argument from historical evidence of rising demand for property rights associated with the rising value of land in several South East Asian countries. Likewise, Ensminger and Rutten (1991) show that the desired system of property rights changed with ensuing economic growth and increasing sedentarization in the case of the Galole Orma of Kenya, as common property near settlements increased in value and the gains from the exclusion of nomadic herds grew. Property rights, then, are created or changed in response to economic forces, as opportunities to gain arise.

In contrast to the economic perspective, legal and political philosophy would emphasize the tendency of property rights to resist and work against change. Thus, one can argue that individuals tend to perceive property rights as rather permanent, and do not constantly perform cost-benefit calculations to determine the presently best allocation of property rights for them.[4] Furthermore, they might value 'property ownership' independently of its direct economic return, because of personal values or status considerations, for instance. In addition, individuals are also guided in their actions by norms and habits, which tend to work against constant and rapid change on the basis of rational calculations.

Finally, legal provisions and normative conceptions in society can prevent a maximization in property rights due to economic calculations. Thus, normative conceptions might prohibit or at least question the possibility of ownership, as in the case of a human being, for instance, or the

[4] In addition, the bounded rationality arguments, of course, apply, as some scholars call it, a boundary of imprecision in human decision making (Windrum and Birchenhall, 1998).

world's oceans. In consequence, distributions of property rights are framed by normative mindsets.

3.2 Property regimes

The literature on environmental resource management traditionally differentiates between different property regimes, i.e. property arrangements characterized by different combinations of property rights, in terms of ownership, access, and withdrawal regulations. The most common categorization of property regimes differentiates between private property, common property, open-access, and state-ownership (Feeny, Berkes, McCay, and Acheson, 1990; Devlin and Grafton, 1998). While these property regimes can exist with respect to any good, we will focus in the following discussion on resources, since this is the focus of the book.

Private property according to this differentiation exists when the exclusive title to the resource is held by individuals or corporations. Accordingly, the respective individual or corporation has control over access to the resource, and is backed in this capacity by the state. Furthermore, the individual or corporation have decision-making capacity with respect to the management of the resource. From a traditional economic perspective, private property is associated with the greatest extent of individual control and the least collective action problems (see below).[5]

Common property as a term has been used or rather abused in the literature because of its vague application (Ostrom and Schlager, 1992). In the conventional categorization of property regimes, it is important to differentiate situations of common property from situations of open-access or common pool resources in general. Common property generally refers to resources for which the exclusive title is in the hands of a group of individuals. This group has control over access to the resource, is frequently backed in this capacity by the state, and has general decision-making capacity over the resource. Common property regimes can exist in that a small, voluntary group owns a resource and can exclude outsiders, but also in cases in which a large, inclusive group with compulsory membership owns the resource (Kasper and Streit, 1998: 186). While legal recognition of joint property rights by government is a sufficient condition for common property, it is not a necessary one. In tribal environments, and even in the developed Western countries with their extensive legal frameworks,

[5] Contrary to many advocates of private property rights claiming that the latter are associated lower costs of governance, however, these lower costs do not imply that a single appropriator from a resource is economically more efficient than a group of appropriators in all instances. Resource ownership by a single appropriator is associated with higher costs of exclusion than ownership by a group (see Eggertsson, 1996).

'property rights' can be based on habit or custom, or codes of conduct other than 'law' in the narrow sense. However, Feeny et al. (1986) argue that the legal status of property rights is extremely important for the environmental outcome.

Open-access refers to situations in which property rights have not been defined, i.e. nobody holds exclusive title to the resource. Accordingly, there is no possibility of access control and exclusion of non-owners, and no regulated decision-making process. From an economic perspective, this situation can be very similar to one in which everybody 'owns' the resource, i.e. the ownership of the large, inclusive group with compulsory membership described above.

Finally, state-ownership generally is used to refer to situations in which the state holds the exclusive title to a resource and controls access to the resource. Frequently, state-owned resources are open to access by the public. Thus, it may seem that state-ownership refers to a situation in which there is limited potential to exclude non-owners. However, this is a function of convention rather than a necessary legal or economic characteristic of state-ownership. Fundamentally, the state can control access to the resource. In situations of state-ownership, decision-making capacity with respect to the resource is, of course, in the hands of the state.

The use of this categorization of property regimes has declined in the literature. The four categories are somewhat artificial since the boundaries between them move depending on whether we look at the property rights distribution from an economic or legal perspective. From an economic perspective, for instance, common property resources and even state-owned resources can fall under forms of private property, a point to which we will return in chapter 5. Likewise, a common property situation with a large compulsory group of membership in which the solving of collective action problems may be almost impossible is not very different from an open-access situation. In addition, the four categories are of limited usefulness in research because most resources exist with different layers of property rights distributions at the same time. While some attributes of a resource may be held as private property, others may exist under common property or open-access regimes. The state, in particular, tends to maintain some rights even in the context of private property. Thus, the differentiation between the four categories of property rights may help as a short-hand in discussions, but it may also create artificial differences where there are none and thereby hinder progress in research. Scholars have therefore moved to different approaches for analyzing property rights distributions in particular and collective action institutions in general. Ostrom (1998, 1999) has developed a broader rule based institutional rational choice approach. Devlin and Grafton (1998) have identified six characteristics of property rights as a basis for differentiation:

exclusivity, divisibility, flexibility, transferability, duration, quality of title. This book, in turn, will transform the four categories of property regimes into two dimensions of property arrangements: the extent of collective action problems among the appropriators and the degree of state intervention (see chapter 5).

3.3 The Tragedy of the Commons

Property rights have been long hailed by economists as a provider of positive incentives for environmental stewardship. The basic idea here is that property rights are important for resource regimes because people tend to take better care of what belongs to them than of the possessions of other people or the collective, and thus should be granted rights to the resources they use. However, there are some problems with this idea which we will discuss after we have laid out the original argument.

From an economic perspective, property rights affect the potential for environmental stewardship by structuring choice sets for consumption and investment decisions. As decision-makers base withdrawal and investment decisions on expectations about returns, property rights define the distribution of incentives decision-makers face when maximizing their utility in the context of scarce resources. Only if decision-makers have the assurance that they can control revenue in the long run, will they have an incentive to maintain the value of a resource and make consumption and investment decisions accordingly.

In terms of *consumption* decisions, the assurance of long-term control induces property owners to limit withdrawal from resources to a sustainable level, so that the resource can continue to provide benefits in the long run.

> The concentration of benefits and costs on owners creates incentives to utilize resources more efficiently (...) The development of private rights permits the owner to economize on the use of those resources from which he has the right to exclude others (Demsetz, 1975a: 31).

Owners of a fishing ground, in this argument, are more likely to harvest just the amount of fish that allows a stable population, capable of renewing itself, to remain, if they can limit the access of other appropriators from the fishing ground. Owners of grazing grounds have an incentive to limit the number of cattle on the pasture to avoid the degradation of the land, if they can be sure that the benefit of 'restraint' will not be predominantly consumed by other cattle owners.

In the absence of control, in contrast, overexploitation and eventually depletion or destruction of natural resources are likely. If decision-makers lack control over long-term costs and benefits in the absence of property rights, because they cannot exclude others from the use of the resource, it may be rational for them to forego long-term benefits in favor of lesser short-term benefits. In other words, if use of the resource by others cannot be prevented, decision-makers can optimize private benefit (in the absence of cooperation) only by increasing their own withdrawal rates above the social optimum, as gains are distributed on a first-come, first-serve basis. As a consequence, fishing grounds are overharvested, hunting grounds depleted, pastures overgrazed, forests destroyed, and air and water severely polluted. Deacon (1994) highlights the link between insecure property rights and deforestation, as well as its consequences:

> For example, insecure ownership induces short harvest rotations on land used to grow trees for timber or biomass for shifting cultivation, and short rotations can in turn cause forest land to degenerate into wasteland. In effect, insecure ownership induces mining of the forest's biomass. Insecure ownership also weakens incentives to develop plantation forests and village wood lots for timber and fuel wood. If such investments were made, timber and fuel provided by planted stands would reduce pressure on natural forests (Deacon, 1994: 2).

Likewise, Dauvergne (1997) demonstrates that insecure property rights due to multiple levels of subcontracting in a politically unstable and unpredictable setting based on corruption and patron-client relationships hasten deforestation in South East Asia. Southgate, Sierra and Brown (1991) find the security of tenure related to deforestation rates in Ecuador. Such consequences can be avoided through cooperation among the appropriators of a resource or government intervention. The former, however, may fail because of collective action problems (see below), while the latter option is associated with problems ranging from efficiency concerns to questions of government willingness and capacity (see chapter 5).

Like consumption, decisions on rate and form of *investment* are a function of control over the benefits provided by a natural resource in the long run. Investing means sacrificing today's consumption for tomorrow's consumption in the expectation of net gain. If the expected net gain is small because of high uncertainty (which translates into the application of a high discount rate to future returns), the potential investor has no motivation to forego today's consumption. The lower the expected returns are, the lower the optimal amount of investment. A natural resource for which no secure property rights exist, in consequence, can suffer from lack of necessary inputs to sustain itself. Appropriators from fresh water sources, for instance, have less incentive to invest in facilities to keep the water clean, if they are unlikely to capture a sufficient share of the return on their investment, because most of the water is withdrawn by other appropriators. Likewise, farmers sharing fields have less incentive to invest in irrigation provisions if they know that rival farmers will capture part of the return on their investment.[6]

If owners do make investments the context of insecure property rights, they have an incentive to invest in inputs that are more easily sold or used elsewhere. They will shy away from high sunk costs, which make disinvestments costly. Furthermore, if the uncertainty of their control results from joint use of an asset with others, they are likely to invest in inputs with higher levels of excludability, i.e. for which monitoring and enforcement costs are lower. Pejovich (1975), for instance, cites the raising of cattle (rather than planting of more profitable almond trees) as an example of a possible change in the form of investment because of insecure property rights.

In sum, according to the original economic argument, the absence of secure property rights can cause environmental (and economic) waste.[7] For natural resources, the subsequent dynamics are expected to lead to unsustainable exploitation rates and underinvestment in necessary inputs. The underlying dynamics are best captured in the literature on collective action and 'the tragedy of the commons.'

A collective action problem identifies situations in which rational individual action leads to a socially Pareto[8]-inferior outcome, i.e. "an outcome which is strictly less preferred by every individual than at least one other outcome" (Taylor, 1987: 19). A collective action problem exists, for instance, if the cooperation of individuals is required for the provision of some socially desirable good, such as environmental quality, but individual

[6] Note, that individuals often do not just calculate their own costs and benefits in such situations though, which is where the economic argument is insufficient. Frequently, individuals will abstain from appropriate investments - even if they would receive a net-benefit - if other individuals have the potential to free-ride, because they conceive such free-riding as unfair. Thus, norms do play a significant role in influencing individual decisions, besides economic incentives.

[7] Economists have identified insecure property rights as causes of "inefficient" behavior, if considered in terms of long-term social objectives, in numerous contexts, including investment rates and types and economic growth (Alesina, Ozler, Roubini, and Swagel, 1991; Barro, 1991; Borner, Brunetti, and Weder, 1995; de Soto, 1989; Ozler and Rodrik, 1992; Persson and Tabellini, 1994).

[8] While this argument relies on the Pareto-criterion, we need to acknowledge that this criterion is not without problems of its own, of course. Two Pareto optimal positions cannot be compared, for instance. Furthermore, it provides no guidance for comparisons of income distributions, and favors the status quo. Still, it remains one of the most prevalent normative standards in economics.

utility maximizing strategies prevent individuals from contributing (a dynamic also captured in the Prisoner's Dilemma[9]).

> The aggregate gains to a group from collective action could greatly exceed the total costs of that action, but it by no means follows from this that the collective action would occur, no matter how rational and intelligent the individuals in that group might be (...) What keeps groups of rational individuals from acting rationally in their collective interest? The most notable thing that prevents this is that, in many situations, the benefits of any collective action go to every individual in some group, whether or not the individual made any contribution to the costs of collective action. In other words, the benefits of collective action are normally indivisible in the sense that, if they are made available to one person in a group they are thereby automatically also supplied to everyone in the group. As is by now widely known, nonpurchasers cannot be excluded from the consumption of the "collective goods" or "public goods" that collective action provides (Olson, 1992: viii).

Any collective action situation is characterized by a divergence between social and private costs and benefits, in other words, externalities, as not all costs of an individual's action are borne by that individual, and not all benefits from an individual's action accrue to that individual. The provision of any amount of a collective good by one individual confers some benefits to other individuals. The degradation of any collective good by one individual affects the value of that good to other individuals. As an individual equates her or his marginal costs and benefits, and ignores the marginal costs and benefits conferred on others, collective failure is the result.

Collective action problems exist whenever property rights are not fully defined, which means they always exist. As the costs of defining rights to every single attribute of a good are prohibitive, the benefits of those attributes for which the costs of defining property rights outweigh the gains are transferred to the public domain. In every transaction involving the exchange of property rights, the costs of obtaining the necessary information about each of the multitude of attributes of a good prevent the complete definition of the rights that are being transferred: "Because transacting is costly, as an economic matter property rights are never fully delineated" (Barzel, 1989: 1). Because of the costs involved in stipulating and monitoring these attributes, parties to an exchange limit negotiations and

[9] The Prisoner's Dilemma offers an alternative way to show how the failure of individuals to cooperate can lead to a Pareto sub-optimal outcome. As defection is the dominant strategy for players in a Prisoner's Dilemma situation, the Nash outcome is mutual defection, which has the lowest social payoff.

delineation of rights to those attributes they consider worth the effort (for which the benefits outweigh the costs, thus satisfying multiple marginal equilibria rather than one). Barzel argues that any attribute for which no stipulations are made and that therefore can be varied becomes a "free attribute," in the context of this discussion "open-access." With these attributes being relinquished to the public domain, anybody can choose to spend resources (monetary or other) on the capture of their benefits.

Thus, while we might tend to think of land over which no property rights have been defined, for instance, as an open-access resource involving collective action problems, even land held as private property is associated with some collective action problems, caused for example by the impact of the specific form of land use on the air or groundwater quality in the area (for which property rights are not defined and enforced). While a better definition of property rights, then, reduces collective action problems, the impossibility of a perfect delineation of property rights implies that collective action problems will always exist.[10] This is especially the case, if we view nature itself as a stakeholder (part of the collective action arena, but unable to speak for itself). In many cases of natural resources management, nature or ecosystem sustainability, i.e. the nature support function of a water resource, for instance, is one of the rival uses to be taken into account when analyzing the collective action problem.

In the context of environmental resources, a complete or partial lack of property rights is frequent and severe collective action problems abound. They reach from global problems, such as global warming, ozone depletion, and questions of biodiversity, to local problems such as the pollution and destruction of fresh water resources. Whether we think of forests as providing oxygen, biodiversity or hunting grounds, rivers and oceans providing fish, water, and a 'sink,' or simply the air allowing us to breathe, all of these goods are associated with costs and benefits for large numbers of people ranging from a village community or a group of appropriators from the same fishing ground to the whole human population.

Environmental collective action problems vary with the degree of excludability of a resource's attributes and the potential to free-ride for non-contributors, as these determine the costs of defining and enforcing property rights. Excludability and rivalness to some extent determine which forms of ownership over the good are meaningful and most likely to occur. Goods

[10] For collective action problems to be completely avoided rather than reduced, contrary to Coase (1960), we do not only need property rights, but also a perfectly functioning market providing for a 'correct' value of those property rights. The question of how to achieve that for many environmental resources is one of the fundamental concerns in the environmental field today, of course, as intergenerational dynamics, scientific uncertainty, and those characteristics of natural resources that make the assignment of property rights difficult to begin with complicate the assignment of monetary values.

that are non-excludable and non-rival are 'pure public goods,' and air pollution abatement (for global pollutants), for instance, belongs in this category. The establishment of private property rights over pure public goods is unfeasible, since individuals cannot be excluded from the consumption of their benefits. Private goods, in contrast, are perfectly excludable and rival, and private ownership over such goods is not only feasible but likely, as individuals trying to capture the full rent from that good will try to exclude others from access through the definition or acquisition of property rights. The decisive determinants (in line with the optimization logic presented above) are the costs of exclusion and the expected benefits to be derived from it: "The logic of economics then suggests that the individual or a group of individuals will try to exclude others from using a good whenever the expected benefits (rent) exceed the cost of policing and enforcing the "claim" to that resource" (Pejovich, 1975: 40). Finally, many goods lie between the two extremes, being non-rival but excludable, or of limited excludability but rival (such as migratory resources). In general, of course, questions of excludability and rivalness are questions of degree rather than extremes.

Thus, forests as hunting grounds, for instance, generally allow the exclusion of outsiders at relatively smaller monitoring and enforcement costs, while the forests' benefits in terms of oxygen provision are almost non-excludable. Excludability, on the other hand, would not matter if the benefits provided by natural resources were not frequently partially rival. In a fishing ground harvested by a group of appropriators, one appropriator's catch depends not only on her or his own effort and input, but also that of the others. The amount of water withdrawn by one farmer from a water source or irrigation system shared with other farmers in the presence of water scarcity affects the water supply of the remaining farmers.

Significantly, with the increasing scarcity of clean air, one of the last natural goods once thought of as abundant and non-rival is now being recognized as partially rival and, through the use of incentive based approaches such as emission trading systems brought closer to the characteristics of a private good. This increasing recognition that natural resources are partially rival, is a function of population and economic growth, technological development, and increasing environmental degradation. As more people consume more goods, as technologies allow us to extract resources increasingly efficiently, and as resources become scarce as a consequence of these developments, we find ourselves in competition for the benefits that remain. Marine fisheries present a very striking example of these developments. Up to the 1950s, the influence of the harvesting of fish by the human race was not considered to threaten the size of the fish population. This changed with technological innovations allowing previously unknown catch sizes to be brought to shore, combined with rising demand.

Collective action problems, thus, reflect a lack of individual control over the long-term use and management of natural resources, and tend to be associated with overuse and exploitation. Hardin[11] (1968) is the one who has drawn the most attention to this dynamic, 'the tragedy of the commons,' in the context of environmental resources. His famous example highlights the results of the divergence between individual and social rationality in the absence of coercion in the case of a communal grazing area.

> Adding together the component partial utilities, the rational herdsman concludes that the only sensible course for him to pursue is to add another animal to his herd. And another, and another (...) But this is the conclusion reached by each and every rational herdsman sharing a commons. Therein is the tragedy. Each man is locked into a system that compels him to increase his herd without limit - in a world that is limited. Ruin is the destination toward which all men rush, each pursuing his own best interest in a society that believes in the freedom of the commons. Freedom in a commons brings ruin to all (Hardin, 1968: 1244).

Besides this well-known example, Hardin reveals the continuing existence of this tragedy of the commons for various other environmental areas. He points out that ranchers leasing national land on the Western (US) ranges demonstrate the respective behavior as they "pressure federal authorities to increase the head count to the point where overgrazing produces erosion and weed dominance" (Hardin, 1968: 1245). Likewise, he recognizes collective action problems determining the fate of the oceans: "Maritime nations still respond automatically to the shibboleth of the 'freedom of the seas.' Professing to believe in the 'inexhaustible resources of the oceans,' they bring species after species of fish and whales close to extinction" (Hardin, 1968: 1245). Finally, he points out that the problem of pollution is the tragedy of the commons in reverse, as the private costs of discharging waste into the commons faced by an individual are smaller than those of purifying the wastes before releasing them.

Even before Hardin, Gordon (1954) analyzed collective action problems associated with a fishing ground. He found that a lack of cooperation among fishermen results in higher expenditures of effort, higher fish landings, and a lower continuing fish population than the optimal level:

[11] Hardin and the 'father' of *Collective Action*, Mancur Olson, were not the first scholars to reveal the dynamics around collective action problems, however, and we should give credit where credit is due. Aristotle already argued that "what is common to the greatest number has the least care bestowed upon it. Everyone thinks chiefly of his own, and hardly of the common interest" (Aristotle, *Politics*, Book II, chapter 3).

> In the sea fisheries, the natural resource is not private property; hence the rent it may yield is not capable of being appropriated by anyone. The individual fisherman has no legal title to a section of ocean bottom. Each fisherman is more or less free to fish wherever he pleases. The result is a pattern of competition among fishermen which culminates in the dissipation of the rent of the intramarginal ground (Gordon, 1954: 130f).

Similar analyses have been conducted by a number of scholars (Sandler, 1992).

The traditional economic argument thus predicts that an absence of (private) property rights will lead to overexploitation of environmental resources and, in consequence, to their degradation or destruction. Hardin was unclear, however, about whether he referred to resources over which no property rights were defined or resources held in common by a group of individuals. While some of his examples such as his discussion of the fate of the oceans refer to open-access situations, the village grazing ground would appear to fall into the category of common property. Subsequent research has focused on the need for better distinctions between these cases and also questioned the generalizability of Hardin's claims.

The tragedy of the commons argument clearly is not without weaknesses, and scholars have raised numerous criticisms against an easy acceptance of its claims. The most fundamental criticism applies to the narrowness of perspective of the analysis. Scholars point out, for instance, that the determinative factors underlying the simple analysis of the Prisoner's Dilemma game, i.e. the isolation of the prisoners and the absence of communication, are absent in many real-life situations in which people depend on a common resource:

> What makes these models so dangerous - when they are used metaphorically as the foundation for policy - is that the constraints that are assumed to be fixed for the purpose of analysis are taken on faith as being fixed in empirical settings, unless external authorities change them. (...) Not all users of natural resources are similarly incapable of changing their constraints (Ostrom, 1990: 6f).

Furthermore, critics find Hardin's (1968) propositions to be too simplistic, as numerous other factors such as the technology of public supply,[12] the payoff structure, length of the 'game,' the modeling of

[12] The technology of publicness captures how individual contributions add to the total public supply achieved. The most common technology identified by Sandler (1992) is that of summation: $Q = \text{sum } q^i$, where q^i is the collective good's provision level of individual I, and n is the group size (or $Q = w^1q^1 + w^2q^2$, in which $0 \leq w^i \leq 1$ for $i = 1,2$, if the substitutability of individuals is imperfect. Alternative technologies are those of the weakest link: $Q = \min (q^1, \ldots.. , q^n)$, and the best-shot technology: $Q = \max (q^1, \ldots ,q^n)$.

cooperation and defection as a continuous variable, and institutional rules have an impact on outcomes, allowing results other than the tragedy of the commons (Sandler, 1992). If a public good is impure in terms of excludability and/or jointly produced private benefits, for instance, Nash behavior need not imply sub-optimality (club goods/clubs). Moreover, in infinitely repeated situations scholars have found that cooperation rather than defection is the rule and endogenous solutions are possible, as parties achieve the socially optimal outcome through strategies such as tit for tat (Olson, 1992). In addition, Sandler argues that the cooperation/defection variable needs to be considered as a continuous variable instead, measuring levels of contribution. Given such an approach, for normal goods the level of provision will increase with group size: "for such models, the total provision of a collective good need not fall with an increase in group size as the increased provision of the entrant more than offsets the aggregate decrease in the collective provision of the existing membership"(Sandler, 1992: 49). Other developments in game theory that have contributed new insights to collective action outcomes are trembling-hand equilibria, work on dynamics and cooperation; and considerations of player types in terms of uncertainty, credibility, and rationality assumptions (minimax, minimize regret, maximize expected utility). Ultimately, the slightest variations in the cost-benefit constellations combined with different technologies of public supply can lead to different collective action outcomes, which makes a reduction of the concept to a few simple propositions impossible (even if we neglect extensions to n-person games).

> Since there are so many different combinations for joining cost structures, the technology of public supply, and tastes, it is not really possible to state general propositions concerning the feasibility for collective action, even in the case of only two people, unless constraints limit these combinations (Sandler, 1992: 44).

Furthermore, the literature on common property resources provides empirical evidence against a necessary link between common pool resources and collective failure (Adger and Luttrell, 2000; Berge and Stenseth, 1998; Bromley, 1992; Charkraborty, 2001; Feeny, 1998; Feeny, Berkes, McKay, and Acheson, 1990; Kissling-Näf and Varone, 2000; Libecap, 1998; Martin, 1992; Ostrom, 1990; Ostrom, Gardner, and Walker, 1994; Schlager and Blomquist, 1998).[13] This literature emphasizes that Hardin's tragedy of the

[13]For collections of case studies see Bromley, Daniel W. (ed.). (1992). *Making the Commons Work. Theory, Practice, and Policy.* San Francisco: ICS Press; Ostrom, Elinor. (1990). *Governing the Commons. The Evolution of Institutions for Collective Action.* Cambridge: Cambridge University Press; National Research Council (1986). Proceedings of the Conference on Common Property Resource Management. Washington, DC: National

commons generally applies to open-access resources, but often is not the outcome for common property resources. Numerous case studies from around the world illustrate that groups of appropriators can jointly and sustainably manage resources. At the same time, a large number of other studies reveal that Hardin's two solutions, state-intervention or privatization often lead to unsustainable outcomes, not to mention social injustices. Adger and Luttrell (2000) show that state appropriation and privatization of wetland resources in Indonesia and Vietnam contributed to unsustainable utilization or conversion. Likewise, Beaumont and Walker (1996) demonstrate that under certain conditions private property does not lead to environmentally desirable outcomes with respect to soil conservation. Finally, recent analyses of resource management situations show that joint ownership can sometimes be more efficient than private ownership due to the costs of exclusion associated with the latter (Eggertson, 1998, 1996).

Scholars working on common property resources have identified a range of factors that foster the ability of joint appropriators from a resource to overcome their collective action problems (Feeny et al., 1990; Ostrom, 1990, 1998; Ostrom et al., 1994). These factors include the homogeneity of the group, the presence of leadership, the ability to exclude outsiders, and most fundamentally the ability to communicate and to learn. Moreover, Ostrom's work, in particular, highlights the need to look at distributions of property rights in more detail: who has which rights to what? As pointed out before, Ostrom differentiates, for instance, between operational rules, collective-choice rules, and constitutional rules in her analyses of common property resource management. In addition, scholars have noted that other institutional arrangements can help individuals overcome collective action problems (Young, 1994).

Those criticisms of the tragedy of the commons argument cannot rule out that the tragedy of the commons can occur. The apparently simple 'rule' that strategies such as tit-for-tat can lead to the social Pareto-optimal outcome as parties learn to cooperate rather than defect, for instance, does not guarantee such a success: "the strategies that support mutual deterrence are frequently so unforgiving and harmful to the individuals who must implement them that their credibility as a viable strategy in field settings is in considerable doubt" (Ostrom, Walker, and Gardner, 1990: 2). Almost any outcome becomes possible in such games including everyone not cooperating with everyone else. Furthermore, while Olson's (1965) proposition that collective failure is

---

Academy Press. Furthermore, a thorough bibliography of Common Property Resource (CPR) studies has been published by Martin, Fenton. (1992). *Common Pool Resources and Collective Action: A Bibliography*. Vol. 2. Bloomington: Indiana University, Workshop in Political Theory and Policy Analysis.

more likely the larger the group in question can no longer be easily accepted, it neither can be easily rejected:

> Even though provision is not inversely related to group size as first hypothesized, sub-optimality and collective failure may, under reasonable scenarios, still worsen with group size in the manner proposed by Olson. An apt example was illustrated (...) with the commons, where each exploiter accounts for one nth of the industry profit" (Sandler, 1992: 194).[14]

Similarly, the applicability of some of the game theoretic extensions is limited in the context of environmental problems. The assumption of repeated games is inappropriate, for instance, when we consider most forms of pollution abatement. As Carraro and Siniscalco (1991) point out, pollution abatement is usually associated with relatively large one-time investments, and the threat of 'defecting' in the future is simply not credible once the investment has been made. At the same time, other extensions underline the continued probability of mutual defection outcomes. Cooperation and tit-for-tat strategies, for instance, become less likely as the number of players increases, and the information set of one player does not include information on whether the other(s) cooperated or not.

In addition, the empirical evidence provided by the CPR case studies concentrates mostly on small scale resources. Some scholars argue that similar dynamics should apply to large scale resources. Because of differences in transaction costs and consequently, payoff structures, however, such a claim is not necessarily convincing. Finally, several worldwide trends such as population growth, migration, and economic growth and development contribute to an increasing instability of successful use and management of common property resources.

On the one side, then, the simple relationship between the exploitation and maintenance of a resource and a lack of private property rights cannot be taken for granted. On the other side, both extensions of the theory of collective action and the empirical evidence on common property resources illustrate that collective action problems and the failure of appropriators to resolve them do exist and are particularly prevalent in many cases in which the group of appropriators is too large and heterogeneous. Furthermore, the enforcement of the exclusion of outsiders, which frequently is required for groups of appropriators to solve internal collective action problems, often is not possible with open-access resources. The literature thus appears to agree that - while common property resources can be managed successfully - open-

[14] The updated version of Olson's proposition then could be: "With identical individuals and symmetric equilibria, an increase in group size worsens sub-optimality when a summation technology applies" (Sandler, op.cit.: 194).

access resources are generally associated with the lowest potential for environmental stewardship.

3.4 Differentiating types of property rights

Some scholars suggest that it is useful to differentiate between different types of property rights, especially as an analytical framework for studying property arrangements in the context of environmental resources. Schlager and Ostrom (1992), for example, develop a categorization of rights that ranges from access and withdrawal rights (use rights) to management rights. They argue that it is important to distinguish between owners, appropriators, claimants, and authorized users of a resource (see Table 3.1).[15] Schlager and Ostrom base this schema on a differentiation between operational level rights of access and withdrawal and more powerful collective choice rights of management, exclusion and alienation. Schlager and Ostrom point out that the five types of rights are independent, but frequently held in the cumulative manner described below (at least with respect to fisheries, to which they apply this conceptual schema).

Table 3.1. Bundles of Rights Associated with Positions[16]

	Owner	Proprietor	Claimant	Authorized User
Access + Withdrawal	X	X	X	X
Management	X	X	X	
Exclusion	X	X		
Alienation	X			

Authorized Users

In Schlager and Ostrom's typology, authorized users are individuals holding rights of access and withdrawal. Sometimes these rights can be transferred to others either temporarily or permanently. Rights of withdrawal and access in turn are defined as the right to enter a defined physical property and the right to obtain the 'products' of the resource. According to Schlager and Ostrom, authorized users have little incentive to invest in efficient resource management, and are likely to seek to gain as much as possible with inefficient outcomes being the probable result.

Claimants

Claimants, in turn, possess the collective choice rights of management in addition to access and withdrawal rights. Thus, they can design operational

[15] In addition, one can talk about the final beneficiaries from the resource, i.e. the consumer of the final product. In the context of our project, however, such a focus appears less important.

[16] Source: Schlager and Ostrom, 1992.

level rights of withdrawal. According to Schlager and Ostrom's typology, claimants cannot, however, design rights of access to resources. In consequence, claimants have some incentives to invest in governance structures for their resources. Given a lack of assurance that rewards for their investments will not be captured by others, however, such investments are highly context dependent. Schlager and Ostrom argue that claimants may invest in resources which are not being utilized by any other group, due for instance to a lack of interest or physical accessibility.

Proprietors

In contrast to claimants, proprietors have the right of exclusion (in addition to operational level rights of access and withdrawal and collective choice rights of management). Thus, proprietors have the authority to decide who may access resources, and how these resources may be utilized. Rights of exclusion provide proprietors with substantial incentives to make "current" investments in resources, as they allow them to be "reasonably assured of being rewarded for incurring the costs of investment" (op. cit.: 257).

Owners

Finally, owners hold the collective choice right of alienation, i.e. the right to sell or lease their collective choice rights, in addition to rights of access and withdrawal, management and exclusion, according to Schlager and Ostrom's schema. Schlager and Ostrom emphasize the importance of rights of alienation for the efficient use of resources, as they provide incentives - if combined with rights of exclusion - for owners to make long-term investments in a resource.[17] Ownership rights thus come closest to what traditional economic analysis considered as private property rights, in that they provide owners with the expectation that they can capture the benefits of long-term investments in a resource. They do not, however, as Schlager and Ostrom point out (and as we saw above), guarantee the survival of a resource in the context of relatively high discount rates.

De jure versus de facto

Schlager and Ostrom further differentiate between de jure and de facto rights. They consider de jure rights to be rights that "may be enforced by a government whose officials explicitly grant such rights to resource users, (...) [so] that they are given lawful recognition by formal, legal instrumentalities" (op. cit.: 254). In contrast, de facto property rights are considered to originate among resources users: "such rights are de facto as

[17] Recall that a sustainable harvest rate is a kind of investment, as it means sacrificing today's consumption for tomorrow's consumption.

long as they are not recognized by government authorities" (ibid.). This differentiation between de facto and de jure property rights highlights the importance of assurance of property rights by the state, as one condition for de jure property rights to effectively exist and to determine (environmental) outcomes. Schlager and Ostrom acknowledge that with respect to a single common pool resource "a conglomeration of de jure and de facto property rights may exist, which overlap, complement, or even conflict with one another" (ibid.).

The strength of Schlager and Ostrom's categorization of types of property rights is that it allows to capture differences in the security of property rights and therefore differences in incentives to manage resources efficiently. Such a conceptual schema of distinguishing among rights provides a superior basis for capturing the harvesting and investment incentives than the traditional differentiation between private property, common property and open-access. Critics of this categorization, however, may point out that the types of rights identified by Schlager and Ostrom may differ greatly within their individual categories. Thus, it will be of great importance for management and exclusion rights, whether they are given for life or until the owner decides differently. Likewise, it will make a substantial difference for ownership rights, if the owner has signed over management and exclusion rights to a proprietor for a substantial period of time, as ownership rights in that case have no bearing on the efficient management of the resource. Schlager and Ostrom themselves demonstrate the necessary precaution in applying this schema with their differentiation between de facto and de jure rights. Indeed, they find that de facto rights of authorized users might lead to a more efficient management of a resource than de jure rights of proprietors.[18] Thus, economists would claim that the legal definition and reach of rights matters little compared to expectations about returns and resulting incentives.

3.5 So what?

What can we learn from these debates? Theoretical arguments and empirical evidence exist for the potential environmental benefit of clearly delineated property rights, the potential of sustainable management of resources by numerous appropriators, the role of governments in contributing to the sustainable or unsustainable management of resources. Open-access resources appear to be most vulnerable with respect to unsustainable management. Moreover, distinctions between different types of property regimes and property rights exist, but may be of limited

[18] An additional problem may arise from the potential difficulty to differentiate between de facto and de jure in traditional societies.

usefulness for this book. As the following chapters will show, a synthesis of these seemingly unrelated and potentially incongruent arguments and evidence can provide the basis for the development of general statements on the environmental implications of property arrangements. As a first step, chapter 4 will focus on the role of assurance of property rights with respect to environmental stewardship. Subsequently, chapter 5 turns of the question of the desirability of government intervention in property arrangements.

Chapter 4

Assurance of property rights and environmental stewardship

If we believe the environmental literature on property rights, then, open-access resources are most susceptible to the tragedy of the commons. Any reduction in open-access resources, therefore, should lead to an improvement in environmental stewardship. This appears to present an opportunity for governments to intervene. Taking a transaction cost based approach, it is possible to show that the provision of assurance of property rights by governments allows a reduction in the extent and number of open-access resources. In sum, in so far as we can assume that open-access resources bear the worst environmental fate, improvements in the assurance of property rights increase the potential for environmental stewardship.

This chapter lays out the argument for positive implications of assurance of property rights in detail. It shows that the provisions of assurance of property rights changes the costs and benefits of defining and enforcing property rights thereby inducing decision-makers to obtain and maintain rights to more resources. The chapter then acknowledges that assurance of property rights is not a sufficient condition for sustainability, thereby highlighting a frequent oversight in the property rights debate. This oversight, the relationship between the "environmental" and "economic" values of a resource, will then provide the basis for the development of the argument of chapter 5.

In the context of this discussion, assurance of property rights refers to institutional support for systems of creation and transfer of ownership rights. Such support derives primarily from the provision of means of recourse for individuals who perceive their rights to have been violated. Assurance, defined in this manner, is generally provided by government.[1] By lowering transaction costs involved in the delineation, protection and transfer of property rights, assurance is instrumental in allowing the process by which property rights are delineated and enforced to become well-developed and reliable. Importantly, assurance needs to be distinguished from the security and enforcement of property rights.

Assurance of property rights differs from the security of property rights in that assurance refers to the general political and legal environment in which those rights exist. In contrast, the security of property rights refers to characteristics of specific rights. Secure rights are those which the owner is not in danger of losing. An individual's rights can be secure, for instance, because of the individual's own ability to protect them.[2] An individual's rights can also be secure, because nobody else wants them. Assurance can enhance the security of individual property rights, or at least lower the costs of achieving a given level of security. The provision of an authority to which individuals can apply for protection of their property rights lowers the chances of these being violated. Further, it lowers the costs for individuals to protect their rights. Thus, while the security of property rights is generally influenced by assurance, assurance is not a necessary condition for secure property rights.[3]

Assurance of property rights differs from the enforcement of property rights in ways similar to the difference between assurance and security. Enforcement, like security, pertains to specific rights.[4] These rights can be enforced in various ways, ranging from private means to the reliance on government made possible by assurance. Assurance provides individuals with a means to achieve the enforcement of their property rights. Yet, assurance is not needed for effective enforcement.

It is important to emphasize that an advocacy of the assurance of property rights does not mean an endorsement of private property regimes. The assurance of property rights is just as important for common property resources and state owned resources as it is for private property. This will become clear in the discussion below.

[1] In traditional tight-knit communities, assurance may be provided by the community as such, in other words by a government of the people.

[2] An illustration of such a case is the existence of private armies hired by drug lords to protect their villas.

[3] As pointed out above, assurance generally facilitates the security of property rights or lowers the costs of achieving that security. If the "rights" in question are not perceived as legitimate by the authority providing assurance, however, the opposite is the case.

[4] After all, enforcement is a necessary condition for security of property rights (monitoring being the other), if these rights are desirable to anybody besides the owner.

4.1 The effect of the assurance of property rights

The assurance of property rights influences environmental stewardship primarily through its impact on transaction costs and expectations. Assurance has the potential to change the extent to which decision-makers define and enforce property rights by lowering these transaction costs and thereby altering the cost-benefit ratio decision-makers face. From the standpoint of an economic rationale, individuals will try to define and enforce their property rights up to the point where the costs of doing so outweigh the benefits. In consequence, property rights will increasingly be defined and a reduction in open-access resources and externalities will occur when the expected gains from internalization rise or when the associated costs fall.[5]

The dynamics depicted here are somewhat opaque, because of the prevalent vagueness in the use of "transaction costs" in economics. For this research, the definition of transaction costs provided by Allen (1991) is probably the most useful:

> Transaction costs are the resources used to establish and maintain property rights. They include the resources used to protect and capture (appropriate without permission) property rights, plus any deadweight costs that result from any potential or real protecting and capturing (Allen, 1991: 3)

[5] According to Coase (1960), internalization will automatically take place if in the presence of specified property rights and absence of transaction costs, parties are allowed to transact. Coase (1960) even argues that in the absence of transaction costs the outcome of the internalization will be efficient and maximize overall returns independent of which party holds the property right/liability. The impact of a change in the rule of liability then would be limited to wealth distribution, which, however, could result in different demands (Demsetz, 1975a). Furthermore, Bromley shows that Coase's argument only applies for producer-producer transactions, as in producer-consumer "income effects become important" and "current endowments and entitlements dominate the outcome" (Bromley, 1978: 51). In addition, Coase's identification of the market forces that would bring private and social costs into equality (without the use of a Pigouvian tax) is limited to a world free of transaction costs: "without transaction costs the contractual stipulations will be so designed that they are consistent with the equimarginal principle" (Cheung, 1975: 438). In the presence of transaction costs, externalities always exist and if private costs of transacting outweigh the gain, externalities will not be internalized and the solution to externality problems may depend on the government. Government intervention, in turn, is usually associated with significant costs itself and does not necessarily increase efficiency. If both the costs of government intervention and the costs of establishing a market (private interaction between individuals) are too high, externalities continue to exist. The government, of course, can choose to ignore the cost-benefit ratio of intervention in terms of efficiency, and choose to intervene for political reasons, independently of the associated costs.

In the context of the definition and enforcement of property rights, two sources of the size of 'transaction costs' exist. One source of transaction costs is the general environment in terms of definition and enforcement of property rights. In a stable environment, where property rights generally are well defined and enforced, information costs are low, and a high level of assurance of property rights exists, transaction costs in general tend to be lower than in an unstable and intransparent environment. The general environment, thus, influences the costs involved in acquiring, transferring, and protecting one's property rights.

The second source of transaction costs is the definition of rights to a specific property's attributes, i.e. of specifying a contract. In terms of the individual good, the costs of defining property rights to a good's attributes are a positive function of their extent and specificity. It would costs more in time and energy if not money to specify the sale of forested land if the number of trees, their individual size and health, and the included plant and animal life were to be specified in the sale, than if they weren't. In contrast, the benefits of defining property rights to the attributes of an individual good are a negative function of their extent and specificity. In terms of the forested land, specifying the property rights to the last bush growing under one of the trees tends to provide less benefit than the rights to the land and trees as such.

The following figure illustrates associated dynamics:

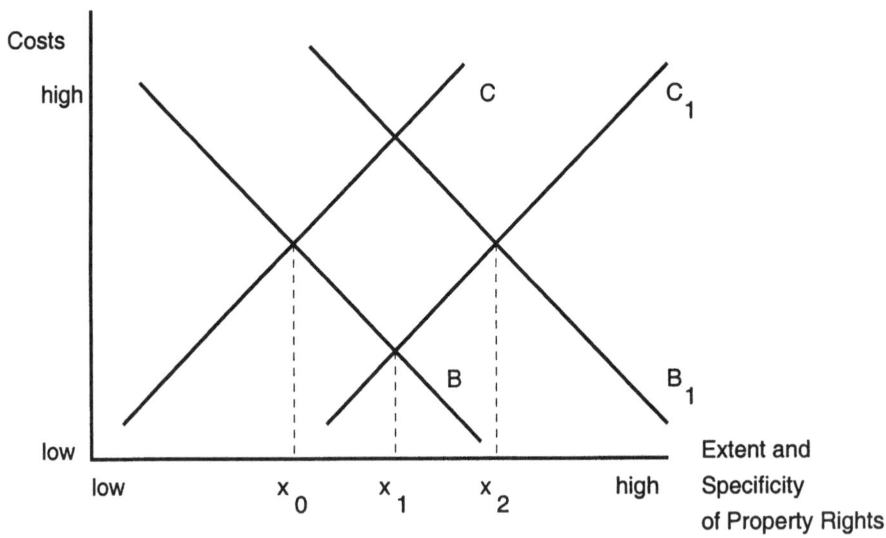

As argued above, the marginal benefits of defining property rights are a negative function of their extent and specificity (B). The marginal costs of defining property rights, in contrast, are a positive function of their extent and specificity (C). Given B and C, rational decision-makers will specify property rights up to the equilibrium position x_0, where the marginal costs of doing so equal the marginal benefits. Anything to the right of x_0 will remain open-access, while property rights will be defined for resources on the left of x_0.

If the general transaction costs decrease because of a better assurance of property rights, that is C moves down to C_1, the new equilibrium will be at x_1. At this point, decision-makers choose to define property rights to more resources, or for more attributes of goods in general, leaving less open-access resources. In addition, increased assurance induces decision-makers to apply a lower discount rate to future revenues from resources, and consequently expected benefits are higher. As B moves to B_1, a new equilibrium x_2 emerges, at which property rights to more resources are defined. Less resources remain as open-access resources. The better the assurance of property rights, the less open-access resources exist and the smaller the likelihood of the tragedy of the commons. While transaction costs and therefore open-access resources will always exist, better government assurance of property rights will reduce them.[6]

The assurance of property rights is similarly important for the protection of property rights, once they have been defined. Owners will choose to protect their rights only as long as the benefits of doing so outweigh the costs. If assurance is low, these costs of protecting one's rights will be higher while the expected benefits will be lower. Those resources for which the expected revenues do not balance the costs involved in protecting the rights to the resource will become open-access resources or at least are likely to be environmentally degraded. Owners who decide not to enforce their property rights to a resource have an incentive to draw as much revenue from the resource as possible before giving it up. Dauvergne (1997) delineates the situation of forest resources in conditions of low assurance, here caused by corruption breeding corruption, in which the owners of forests or licenses to harvesting forests have every incentive to withdraw as much benefit as possible in a short time. Given their limited ability to receive governmental support for their property rights to timber, due to the

[6] Specifically, open-access resources will exist as long as the increase in costs associated with the definition and enforcement of property rights to additional attributes of a resource outweighs the decrease in transaction costs associated with the better definition of property rights overall. In other words, if the cost curve associated with rights to the attributes of an individual good is steeper than the curve associated with the overall relationship between property rights and transaction costs, a rational decision-maker will leave rights to at least some of the attributes of a good undefined.
In addition, some externalities will be particularly difficult to internalize at 'true' prices, even if property rights are defined, because markets for them might be too 'thin' or external diseconomies might be present (Dasgupta and Mäler, 1994: 28).

corruption of the system as well as the questionable legitimacy of the system as well as their rights, these actors will not concern themselves with sustainable harvesting methods.

The situation is even clearer if we think of governments that are not incapable of protecting property rights, but are predatory and unwilling to do so. Property rights appear to be perceived as especially insecure when they are threatened by government itself. In many developing countries, frequent appropriations of private property by the respective elites in power have left deep marks in the economic and social landscape. In their analysis of obstacles to investment and economic growth in Nicaragua, Borner, Brunetti, and Weder emphasize the impact of an inconsistent and unreliable government by contrasting Nicaragua with the former German Democratic Republic (GDR):

> The discretionary power of the state and its arbitrary enforcement of rules is a more profound source of investor anxiety than the narrow issue of specific property rights (...) The crucial difference between Nicaragua and the former GDR is that, in the latter case the deeper problem of institutional instability per se was never present (...) From the first day of the unification, the East German entrepreneur - and, in fact, any other investor - knew with certainty that his investments in this region would be protected by exactly the same institutions as protected those in West Germany" (Borner, Brunetti, and Weder, 1995: 53).

The higher the chances, then, that resources will be appropriated by the government, the higher the discount rate owners will apply to future revenues from the resource. As a consequence, they will have every incentive to increase current revenues relative to future ones, which implies the rapid depletion of resources.

> When governments are volatile or predatory the individual's incentive to invest is reduced. Instability in the laws and institutions that govern ownership subjects the future return to an investment to confiscation risk - by outright seizure of assets, failure of government to enforce laws of property, or opportunistic taxation. Rule by an elite group rather than by laws can have similar effect, since the individual's claim to property depends on remaining in favor with the ruling group and on the group's hold on power. Weakened ownership, in turn, dampens the incentive to accumulate and conserve capital of all sorts, including forests, mineral reserves, and ordinary produced capital (Deacon, 1994: 1).

The absence of government assurance of property rights is especially devastating in the context of state-owned natural resources, because of the concurrent lack of any private incentives to protect the property rights. In the absence of enforcement, state-owned property automatically becomes *de facto* an open-access resource (independent of any cost-benefit calculations which could prevent the deterioration of private or common property to open-access) with the associated collective action problems and resulting overexploitation.

> While most of the world's forests are nominally owned by governments, they are often treated as free access resources when government lacks the ability to enforce controls on their use. This is evident in Latin America and elsewhere by the colonization of national parks and government forest reserves by squatters (Deacon, 1994: 3).

As a consequence, the nationalization of natural resources in some developing countries has had devastating environmental consequences (Adger and Luttrell, 2000; Ostrom, 1990). Rather than placing the previous common property resources under the control of unitary decision-makers, nationalization in the absence of the ability of governments to assure their own property rights increased the number of decision-makers and environmental degradation by turning limited access into open access.

> Alarmed by deforestation, the government nationalized forests in 1957, converting what were often communal forests into de jure state property. But the result more closely approximated the creation of de facto open access. Villagers whose control of nearby forests had been removed often succumbed to the incentives of law of capture. Deforestation accelerated instead of decelerated (Feeny et al., 1990: 8, on Nepal[7]).

Assurance of property rights is not just important for private property, common property, or state-owned property, however. General assurance of property rights can indirectly improve environmental outcomes even under conditions of open-access. The former dynamic is partly the outcome of an established framework of stable and respected property rights, and an internalization of the notion of rights and responsibilities, which can induce individuals to forego strategies maximizing material net-gain.[8]

> Institutions are rules, enforcement characteristics of rules, and norms of behavior that structure repeated human interaction. Hence, they limit and

[7] For an interesting comparative study of today's fate of state owned common property forests in Nepal and common property forests see Chakraborty (2001).

[8] This "norms of conduct" argument has often been used by critics of the collective action argument to show that secure property rights are not necessary for a sustainable outcome. What these critics did not recognize, however, is that the argument can also support the positive influence of secure property rights in cases in which they cannot be defined or enforced.

> define the choice set of neoclassical theory (...) To the degree that individuals believe in the rules, contracts, property rights, etc. of a society, they will be willing to forego opportunities to cheat, steal or engage in opportunistic behavior (...) The function of institutions is to provide certainty in human interaction, and this is accomplished by the inherent features of rules and norms (...) It is norms of behavior, however, that probably provide the most important sources of stability in human interaction (North, 1989: 1324).

North, of course, also recognizes the limits of guidance in decision-making provided by rules and norms of behavior, especially when it comes to utility maximization in the context of changing costs and benefits: "Empirical evidence suggest the price we are willing to pay for our convictions is a negatively sloped function, so that ideological attitudes are less important as the price increases; (…) " (North, 1989: 1322).

Besides this influence of norms, however, the perceived improvement of environmental outcomes may also result from the frequent influence of the conditions of 'neighboring' open-access resources on private property. The quality of the air or the pollution of a lake impact the utility the owner of a house on the lakeshore will draw from her or his property. In the absence of secure property rights, the property itself is relatively less valuable and therefore justifies less investment of effort in obtaining a clean environment. If secure property rights assure the value of the property, however, the owner stands to gain long-term benefit from pollution abatement. The benefits of an investment of material and other efforts in pursuing improved quality of the respective open-access resource (either through the private sector or through the political sector via political demands for environmental measures) are therefore more likely to outweigh the costs. Of course, the quality of the resource that the owner of adjacent property would be interested in does not necessarily have to be its ecological health. The lake, for instance, has to look nice, but does not need to be a healthy ecosystem. In general, however, we can argue that the change will primarily be in the direction of a cleaner rather than a dirtier environment. Thus, an environment of secure property rights will indirectly also benefit those environmental goods over which property rights traditionally have not been specified, such as the air or water resources. While such conditions will not generally lead to dramatic changes (partly because of collective action problems), assurance of property rights can nevertheless lead to environmental improvements in open-access resources.

In sum, assurance of property rights is positively related to the extent and specificity of property rights' definition and enforcement by individuals. If resources for which no property rights are defined and/or enforced are most vulnerable to the tragedy of the commons, the assurance of property rights is of fundamental importance for environmental stewardship. Overall, then, the willingness and capability of government to assure property rights is paramount for guiding consumption and investment decisions made by appropriators from natural resources in the interest of environmental stewardship. In the absence of government assurance, property ownership becomes a question of power rather than right, and thereby subject to arbitrary forces and instability with potentially dramatic environmental (and economic) effects. Without consistently enforced laws, "people are caught in a prisoner's dilemma", where "cheating is the dominant strategy" (Borner, Brunetti, and Weder, 1995: 45). The assurance of property rights - as much a political as an economic variable - is of fundamental importance for environmental stewardship.

4.2 Determinants of assurance

On what does the assurance of property rights by government depend? Two factors are determinative: willingness and capacity.[9] The former is a function of government preferences. In general, the willingness of government to supply processes for the definition and enforcement of property rights has been related to the theory of the state. Government or, initially, an individual or group of individuals in leadership position extend protective services to individuals in their domain in exchange for a share of the individuals' income. Government, thus, can be viewed as a firm that produces and sells protection in exchange for revenue (Pejovich and Furubotn, 1975). According to this logic, then, the willingness of government to provide assurance for property rights (like the willingness of government to define or alter property rights) depends on a cost-benefit analysis, i.e. on the "relationship between the benefits to the authority from granting new or modifying the existing property rights assignments and the costs of protecting them" (Pejovich, 1975: 43). Feeny finds empirical support for such a view in cases of land reform in South East Asia in which "a major motivating factor of the government in the creation and administration of the land-rights system was the desire to collect land tax

[9] Any political scientist will eagerly point to political legitimacy as a precursor of assurance, which then opens up a series of related questions, such as what creates political legitimacy. If we wanted to address the issue of political legitimacy then, we would need to trace the question back further. While an extremely interesting endeavor, this question will therefore not be pursued further in this analysis. Rather I intersect the causal chain at the level of political assurance, and move from there towards the political outcome in the form of environmental quality.

revenues" (Feeny, 1988: 295).[10] Public choice advocates in turn view the individual politician as selling 'rights' and protection in exchange for political support. Finally, as North (1989) points out, the provision of institutional change is also affected by 'ideology' and 'conventional wisdom,' as they define our view of how the world should and does operate.

Beyond the provision of the institution of property rights, scholars frequently view the creation of property rights and determination of their contents by government, rather than based on private incentives, with doubt. North argues that political systems have the inherent tendency to produce inefficient but measurable property rights in order to gain higher revenues (North, 1989). Likewise, Sanderson (1995) highlights that the definition and imposition of intellectual property rights to biota under the biodiversity convention, which he considers incapable of achieving the preservation objective, is the *political* result of the success of development and commercial interests. Furthermore, North and Thomas (1973) claim that the protection of property rights by governments during the Industrial Revolution in Europe was successful and supportive of economic growth, because authorities recognized the efficiency of the established trade practices by the private sector and supported them with law rather than attempting to create a new contract system. Barzel summarizes the role of government and the benefits of its fulfilment of that role, while also highlighting the limits to government action and the importance of the economic context:

> Rights are created in presence of state authority which has a comparative advantage over private individuals in the use of violence and which tends to discourage its private use. When a state authority is in place, the role of allocation devices other than violence is greatly enhanced (...) Governments participate in the definition and enforcement of property rights, but 'economic' individuals are responsible for most, and have a comparative advantage over the government in many of these activities and actually undertake most of them (Barzel, 1989: 63ff).

In turn, the capacity of government to assure property rights is a function of the overall resources available to government, the efficiency of their utilization, and competing demands for government resources. Resources include material resources from taxes and mineral resources, as well as 'human' resources in form of government support from the population as well as from powerful forces in the state such as the military or economic

[10] Furthermore, Feeny links the supply to the cost of institutional innovation, which depends on the stock of existing knowledge, as well as on private benefits and costs of providing the change that agents in the position to provide the change face (1988: 273f).

elites. Such resources are limited, and need to be spread between multiple goals. In the long term, the overall amount of resources available to government depends on a country's endowment with natural (and human) resources, as well as the perceived legitimacy of the government and its policies. If a government lacks widespread support among its populace, its activities become more costly. Not only will government access to resources be inhibited, governments will also need more resources to achieve a given policy objective. The amount of resources available for a given policy objective, in turn, depends on competing demands for resources. In the case of external political threats, for instance, fewer resources will be available for domestic policy objectives. These determinants of resources available to government overall and for specific policy objectives highlight limits to the governmental capacity to provide assurance.

The ability of government to provide assurance for property rights depends, of course, also on the nature of the assets, and the costs involved in monitoring and enforcement. Technological changes allowing new forms of measurement might change those costs and thereby increase a government's capacity to provide not only enforcement but also assurance of property rights with respect to a specific asset:

> Rules (and their enforcement) are constrained by the costliness of measuring the characteristics or attributes of what constitutes rule compliance or violation. Hence, the technology of measurement of all the dimensions (sight, sound, taste, etc.) of the human senses has played a critical role in our ability to define property rights and other types of rules (...) The relationship between the benefits derived from rule specification and the costs of measurement not only has been critical in the history of property rights (common property vs. private property) but is at the heart of many of the issues related to the structure and effectiveness of enforcement (North, 1989: 1321).

If measurement were costless and feasible in all instances, enforcement and assurance would be less of a problem. As a consequence of monitoring and enforcement costs, governments in developing countries sometimes lack the capacity necessary to protect legally defined property rights; "in virtually all cases the legal provisions 'exceeded' administrative practice in the degree of sophistication and precision of the land rights" (Feeny, 1988: 295).

Government capacity has recently received a lot of attention in the political science literature. Scholars have tried to explain and assess government capacity and link it to political, economic, and social outcomes at the local, national, and international level. In International Relations, Arbetman and Kugler (1997) have had some success in estimating the political capacity of governments using macro-level economic, fiscal, and employment data. They and others have applied their measure to explain and predict the outcome of political and economic phenomena from wars to inflation. Jackman (1996) has struggled with defining political capacity on the basis of legitimacy and authority for the field of comparative politics. In the environmental politics literature, Press (1997) and Orth (1997) are among the scholars developing capacity measures. While political scientists have yet to agree on the definition and measurement of political capacity, then, research by this group of scholars increasingly emphasizes the crucial implications of political capacity for assessing the role of the state.

Thus, we can recognize the link between periods of political instability, and lack of management of state property, assurance of (private) property rights and environmental degradation.[11] In support of such a link, Deacon (1994), for instance, finds a positive relationship between political instability and resource degradation, when studying historical deforestation in the Mediterranean region as well as recent deforestation around the globe. In his analysis, Deacon relies on Thirgood's findings on the historical relationship between the deterioration of forests in the Mediterranean region and political unrest.

> In the Mediterranean environment there is a clear relationship between the security that accompanies stable government and good husbandry of the land. Disruption of settled government has almost inevitably led to an increase in pastoralism. This has been so from the breakdown of the Roman Empire, when the nomads were no longer kept beyond the frontiers, through the Moslem invasions of the early Middle Ages, and the upheavals of the Crusades, down to the disruptions of the Second World War and its aftermath (Thirgood, 1981: 58).

In contrast, evidence of forest management in stable periods is even found in ancient Greece and the Ptolemaic period (332-330 BC) in Egypt (Deacon, 1994).

As a result of the variability of factors determining the willingness and capacity of governments to enforce property rights, significant variations in the assurance of property rights exist across the world. These variations are highlighted by North (1989) among others, who finds "vast differences in the relative certainty and effectiveness of contract enforcement, temporally over the past five centuries in the Western world, and more currently between modern Western and Third World countries" (ibid.:1323).

The assurance of property rights on a constant and fair basis does not have to be, but generally is, the job of government (we can think of the

[11] Furthermore, I am ignoring here the environmental degradation directly caused by warfare, which is significant.

assurance of property rights both as a public good and in relation to economies of scale). In small communities with dense social networks and in the context of natural resources associated with limited monitoring and enforcement costs, enforcement can more easily be provided from inside the community. However, in our increasingly interdependent world (and given the high costs of monitoring and enforcement associated with many types of natural resources), the capacities required for monitoring and enforcement are often too large for private provision, and third-level collective action problems too big. Especially with the increasing extension of government legislation in developing countries into remote areas, and the breakdown of local organization of common property use because of population pressure and the increasing scarcity of resources (among others), the capacity of government to enforce property rights is extremely important for sustainable resource use. However, scholars have pointed out additional influences on enforcement. Levi (1988), for instance, argues that in large societies 'Quasi-Voluntary Compliance' is necessary to aid government in enforcement.

4.3 Assurance of property rights - a Panacea?

According to this logic, then, assurance of property rights by government can significantly benefit environmental stewardship. By raising the benefits and lowering the costs of the definition of property rights, government has a direct impact on the environment. The better assured property rights are, the higher people's incentives to extend the definition of property rights and to thereby reduce the amount of open-access resources and the associated collective action problems.

If assurance of property rights indeed uniformly improves environmental stewardship, its use as a policy instrument will provide governments with the opportunity to improve environmental quality at a given level of development. Per capita income levels are difficult to influence. Furthermore, no government would easily set out to negatively influence them, even if rising per capita incomes mean a deterioration in environmental quality. Rather than having to accept environmental dynamics as determined by the level of per capita incomes in the country, governments would be able to take an active stance.

Furthermore, assurance of property rights would not be just an instrument to improve environmental quality, but also an instrument with extremely wide ranging effects. Rather than having to address each environmental problem and indicator by itself, governments would be able to utilize secure property rights as a single measure to influence environmental quality across the board. Given the increasing number of environmental concerns and the limited funds and administrative capabilities, which rarely allow the adequate focus on each one, such a general tool would be immensely valuable.

> In an era of rising interdependencies among differentiable human activities, institutional arrangements should be created with a broad enough scope to encompass activities that impinge on each other. Thus, the days of single-species management systems are essentially over (...) need to strike balance between inclusiveness and workability (Young, 1994: 75).

Moreover, assurance of property rights would have the advantage of being associated with desirable 'side effects.' It would not hinder economic growth as environmental policy instruments are sometimes accused of doing, but aid growth (or in fact provide a precondition for economic growth) according to mainstream economic analysis. Similarly, the assurance of property rights would not take funds away from economic policy for the sake of the environment, but help both.

It is important to keep in mind, of course, that developing countries sometimes lack even the resources needed to provide assurance of property rights. The overhead required to administer, adjudicate, and enforce law in Western societies can frequently not be afforded by developing countries: "the transaction costs of well-defined and enforced private property typical of the West may simply be too great for a subsistence economy to bear" (Runge, 1992: 19). Shleifer (1995) raises similar concerns complaining that the literature advocating private property rights, is "not very specific in (...) explaining how to establish well-defined property rights" (ibid.: 93).[12] Yet, while the process of establishing the institutional foundation of assurance of property rights is difficult and needs further investigation, the cause may be worthwhile to make the effort. Furthermore, the necessary resources for this task should be obtained more easily than additional resources for complex environmental policies. Moreover, private property institutions with local enforcement, backed rather than conducted by government, have proven successful in many cases, especially in developing countries. Placing the control and ownership incentives in local hands thus does not necessarily require an impossible amount of resources.

There may be another concern about the influence of assurance of property rights on environmental stewardship, however. Differences between

[12] Even in developed countries with sufficient resources for the enforcement of property rights, the change in the definition of the latter necessary for the environment might not be politically feasible. Schelbert-Syfrig and Zimmermann (1988), for instance, argue that the vested interests in the Swiss timber industry are too powerful to allow a change in property rights which would bring about an internalization of the costs and benefits associated with the forest industry and forests as such.

two countries in the assurance of property rights might allow one to export environmental degradation or limit environmental stewardship to local or national problems. A high level of assurance of property rights in the Western European countries or the United States, for instance, will have less effect on protecting biodiversity in the Amazon region than on local environmental conditions such as the local/regional air quality in these nations. The positive relationship between the assurance of property rights and environmental quality, therefore, should be stronger for environmental indicators for which cause and effect are in close spatial proximity, than for those for which cause and effect are spread apart.

4.4 Limits to the positive environmental effects of assurance

There are even bigger limits to the ability of assurance of property rights to improve environmental stewardship, which need to be acknowledged here. As shown, the assurance of property rights can be theoretically linked to a reduction in open-access resources. Open-access resources, in turn, allegedly are most susceptible to environmental degradation. Yet, this is not always true. Sometimes the uncertainty of long-term benefits associated with open-access resources will render an exploitation of the resource economically unattractive, while a private property regime would lead to degradation and depletion. And if the assumption of this chapter's argument with respect to open-access resources is not always correct, neither can the deduction that assurance of property rights uniformly improves environmental stewardship be.

In fact, the problems the alleged vulnerability of open-access resources and therefore the assurance argument face are reflected in fundamental oversights in the popular property rights debate. In this debate (that has been taking place in the US in particular) both critics and advocates of (private) property rights have overlooked important caveats to their claims. Most fundamentally, the advocates do not take into account that the economic value of a resource, which most property owners tend to maximize in the long run, is not necessarily equal to its environmental value. On the other side, the "many observers who have come to see existing structures of private property rights as a major source of environmental problems" (Young, 1994: 5) fail to acknowledge the situations in which private property rights lead to the environmentally and economically most efficient outcomes. They also tend to overlook the limited capability of many governments to enforce their own property rights and manage their resources, which severely limits state ownership as an alternative strategy.

While both sides are right in their claims in some cases (which, of course, they use as empirical support for their arguments), they are wrong in others.

Likewise, the relationship between assurance of property rights and environmental stewardship is ambiguous. On the one side, the above logic shows how assurance of property rights can increase the potential for environmental stewardship through inducing a reduction in the number and extent of environmentally vulnerable open-access resources. On the other side, it is not difficult to find empirical evidence that assurance of property rights might sometimes be associated with the inability to achieve environmentally desirable outcomes.

The property rights and wise use movements in the United States have shown that a high level of assurance of property rights can sometimes hinder government in pursuing environmental objectives. Courts have ruled against government on the basis that "taking without compensation" is unconstitutional and required substantial payments of indemnities to property owners for the imposition of use restraints on private property for the benefit of the environment.[13] While in some cases, governmental bodies have developed innovative policy tools, such as tradable development credits, to circumvent the problem, in other cases property owners have been successful in preventing government from protecting the environment.

The issue of taking has inspired a continuing battle between environmentalists and members of the property rights and wise use movements in the United States. It is based on the fifth Amendment which specifies: "(...) nor shall private property be taken for public use without just compensation." Historically, this amendment derived from the American experience under British rule and was meant as a guard against the crown, i.e. to protect the individual from extortion by an undemocratic and 'unjust' government. Today, environmentalists find the amendment being used for what they consider an opposite purpose: as an argument against the collective good, preventing government from pursuing environmental preservation for future generations. While the protection of private property against government intervention is a long-standing constitutional provision, environmentalists emphasize the difference in context between its original conception and its current use. The property rights and wise use movements use the fifth amendment, for example, in attempts to prevent the government from open-space preservation or the protection of wildlife through the endangered species act.

The property rights and wise use movements defend their action by either portraying property rights as 'inalienable rights' or explain the need to protect private property against government intervention on a utilitarian

[13] For a description of the change in state legislation that is taking place because of increasing numbers of court rulings in favor of property owners, see Emerson and Wise (1995).

basis. Individuals sharing the latter perspective argue that a frequent loss in the value of private property due to government action would lead to an alienation of American citizens and a loss in commitment to the republican idea of self-government (Span, 1996).

The questions and problems involved in this issue are many. Does the bundle of rights of property owners necessarily include the right to develop their property, for instance, as members of the property rights and wise use movements claim? If so, is the government's interference with that right preventing the property owner from inflicting harm on the public or demanding a contribution to the public good (the latter one requiring compensation, the former not, according to Span (1996))? None of these and the numerous other associated questions can be answered in this analysis, unfortunately, as each of them requires an analysis in themselves. For the purposes of this study, it is important to note that given certain conditions high assurance of property rights may work to decrease the potential for environmental stewardship. The crucial task is, then, to identify these conditions.

Consider the following examples:

1. A privately owned wetland on which development is encroaching. With the encroaching development the value of the land rises, and it becomes increasingly economically attractive for the owner to sell the property to a developer. The developer, of course, will drain the wetland, and build residential and/or commercial developments on the area.

In this case, an increase in assurance would facilitate environmental degradation. As pointed out above, assurance lowers transaction costs, and thereby increases the expected economic benefit from property, while lowering the costs of defining and enforcing property rights. For the case considered, higher levels of assurance imply that the acquisition of property rights to the wetland is even more attractive to developers. As a consequence, the latter are willing to pay a higher price, which increases the incentives for the original owners to sell the land. In sum, higher levels of assurance facilitate and encourage development of the wetland for residential and/or commercial use. Rather than increasing the potential for environmental stewardship, higher levels of assurance actually lower it.

Consider another example:

2. A fishing pond in the middle of nowhere. The land and fishing pond have no other potential economic use than as a source of fish for their owner.

In this second case, an increase in assurance would increase the incentives for environmental stewardship. If the owner were to fear the loss of her or his property rights, she or he might be inclined to catch all of the fish before that is the case. As long as such fears do not exist, the best the owner of the fishing pond can do is to limit the catch to a size that allows a fish population capable of renewing itself to remain.

These two examples clearly show that assurance of property rights can work for and against environmental stewardship. Under which conditions is one or the other the case? What is the difference between the two situations? In other words, what would a more general theory about the relationship between assurance of property rights and environmental quality look like?

The primary cause of the apparent contradiction in environmental consequences of assurance is the fact that the environmental value of a natural resource might differ from its economic value. 'Economic value,' in this context, refers to the *maximum long-term economic value to be obtained from any use of the resource*. In contrast, 'environmental value' refers to the *economic value that can be obtained from the environmentally most desirable use of the resource*.

The most fundamental criticism for the argument against open-access in environmental management, therefore, results from its implicit assumptions about economic and environmental values of resources. This criticism is particularly powerful as it stays within the economic logic adopted by the proponents of the 'tragedy of the commons' argument and even those merely arguing for the particular vulnerability of open-access resources, and therefore challenges them *on their own turf*. After all, under the assumption of homo oeconomicus, the control over natural resources provided by property rights will only lead to an environmentally superior management of the resource, if the greatest economic benefit to be derived from the resource results from its environmentally desirable use. The environmental and economic values of resources, if defined in this manner, are likely to diverge to varying degrees for any resource. This has fundamental implications for the benefits of assurance of property rights for the environment.

The arguments and examples presented above suggest that assurance of property rights is most desirable and helpful for environmental stewardship with respect to resources whose utilization is dependent on their environmental health. If property owners can achieve substantially higher economic benefits from resource use that is opposite to its environmental benefits, assurance may have a negative impact on environmental stewardship. Thus, the deciding factor that underlies the impact of assurance on environmental stewardship is the difference between the 'environmental' and the 'economic' values of a resource. If the two values are approximately equal, assurance of property rights has the potential to increase environmental stewardship. If the economic value is substantially bigger than the environmental value, however, assurance is likely to weaken environmental stewardship.[14]

Let us call the difference between these two values the '*e-e gap*.' The *e-e gap* is important for the environmental desirability of property rights and arrangements because it determines the implications of the maximization of expected (economic) utility for environmental stewardship. If economic and environmental values of a resource are close, the decision-maker's maximization of expected utility can imply a maximization of environmental stewardship, and therefore the most efficient property regime in economic terms may also be the most desirable one environmentally, i.e. lead to the least environmental overexploitation or degradation. If, however, the difference between the two values is large, the maximization of expected utility is likely to result in the maximization of environmental degradation, and therefore the economically most desirable property regime will not be the environmentally most desirable one.

The *e-e gap* is a function of market prices, as these determine both the maximum economic value derivable from a resource, as well as the economic return on its environmentally desirable use. Besides market prices, characteristics of a resource, such as annual growth rates relative to the current interest rate influence the *e-e gap*. As the interest rate indicates the prevailing return on investment, it highlights the opportunity costs involved in a restraint on harvest. Clark (1973) has shown for whales, and Berkes (1996) and Beaumont and Walker (1996) convincingly argue that the same dynamic applies to redwoods and soil conservation respectively, that for slow growing resources economic rationality does not lead to sustainable harvest rates, due to the role of opportunity costs.

The *e-e gap* thus highlights the limits of the argument that open-access resources are the most vulnerable environmental resources and therefore of the ability of assurance of property rights to improve environmental stewardship. Recall that advocates of the property rights based approach argue that the lack of individual control and responsibility provided by open-access resources leads to high levels of collective action problems and consequently economic and environmental waste. As the above discussion shows, economic and environmental waste (or if put in positive terms, economic efficiency and environmental desirability) are not always closely correlated. Only if the *e-e gap* is small, does greater control imply superior environmental stewardship. In cases in which the *e-e gap* is large, greater control implies a greater potential for an economically efficient environmental degradation of the resource. Note that this argument only applies if we conceive of the owner as 'homo oeconomicus' rather than

[14] The underlying question then becomes how to make the environmentally most desirable use economically more rewarding.

'homo oecologicus.' This assumption cannot always be made, of course, especially when it comes to environmental values. Certainly, a range of other factors influences individuals in their decisions. With respect to the use and transfer of property rights to natural resources, however, economic rationality appears to be the most powerful force in most decisions in the Western developed countries and many of the increasingly Westernized parts of the world. Experience shows that there are limits to the guidance in decision making provided by personal convictions and norms of behavior (North, 1989).

While being an effective tool to provide a higher potential for environmental stewardship in many instances, the assurance of property rights, thus, cannot guarantee an environmentally sustainable outcome as long as environmental and economic values of resources significantly diverge. This divergence is frequently the case, as long as environmental costs and benefits are not fully internalized. Only if incentives are right, i.e. environmental goods are imputed their 'true' value, does a decrease in the number and extent of open-access resources and therefore the assurance of property rights unequivocally lead to improved environmental sustainability. In other words, the valuation of natural assets would have to be based on the principles of sustainable development.

Sustainable development, of course, means that the interest and welfare of future generation is taken into account, and that the current generation does not leave future generations with less than what it had itself in terms of natural assets. In contrast, conventional economic valuation is based on maximizing the interests of the current generation. Specifically, societies value costs and benefits in the near future more than those in the distant future, and after a certain point in time, the value we attach to costs and benefits is negligible. Thus, while it is important to note that the application of property rights to natural resources does not necessarily have to mean a neglect of questions of interdependence and intergenerational justice, irreversibility, and uncertainty; in our current economic calculations it usually does.

4.5. Conclusion

Assurance of property rights has the potential to improve environmental stewardship by reducing the number and extent of open access resources. Improved assurance of property rights lowers the costs and increases the benefits of defining and maintaining property rights to natural resource and thereby induces individuals to leave less resources or attributes of resources in the public domain. Assurance of property rights is not a sufficient condition for superior environmental stewardship, however. As the above discussion shows, the difference between the 'economic' and 'environmental' values of a resource needs to be considered. If a resource's economic value is substantially higher than its environmental value, the economic incentives are aligned against an environmentally desirable use of the property. Under those circumstances, assurance of property rights aids the environmentally non-desirable use and has the potential to hasten environmental degradation. In contrast, if the environmental and economic values are aligned, assurance of property rights enhances the potential for environmental stewardship.

Given those contradictory implications should governments provide a high level of assurance of property rights? The answer is yes. Assurance of property rights has sufficient environmental benefits to be politically desirable. In addition, assurance of property rights has economic and social benefits, which makes it politically popular. However, political decision-makers need to keep in mind that assurance of property rights alone does not solve environmental problems, especially in combination with a lack of internalization of environmental costs and benefits. Thus, efforts to improve the assurance of property rights need to be accompanied by additional environmental measures.

These results have far reaching implications for the property rights debate. They show that general arguments about the relationship between property rights and environmental quality are frequently incomplete. Importantly, this conclusion applies to the whole range of facets of the relationship between property rights and the environment. The relationship between a resource's economic and environmental values is important with respect to the environmental implications of assurance, security, and form of property rights. Similar to assurance, the security of property rights only has positive implications for environmental stewardship, if the property will be used in an environmentally desirable way, i.e. if the economic value of the resource is not substantially higher than its environmental value. In the opposite case, security facilitates environmental degradation. With respect to the environmental implications of the form of property right, i.e. property regimes, the story is more complex but fundamentally similar.

This discussion implies that the environmental effects of property arrangements have to be studied in the context of the economic and environmental values of resources. To date, this factor is not being considered to a sufficient extent. As pointed out above, the relationship between a resources environmental and economic values is a function of the degree of internalization of environmental costs and benefits. The book thus joins the chorus of environmental policy, politics, and economics arguments emphasizing that internalization of environmental costs and benefits is extremely important. Individuals need to face the correct costs and benefits when defining and enforcing property rights to natural resources as with any other economic decisions.

Chapter 5

The environmental desirability of government intervention

Besides providing assurance of property rights as discussed in the previous chapter, governments can influence the environmental implications of property rights through the intervention in 'private' property rights on behalf of the environment. Private, in this context, means property rights not owned by government, i.e. includes both 'private property regimes' and 'common property regimes' as well as open-access. This chapter, therefore, focuses on the question to what extent government intervention in private property rights is desirable from an environmental point of view.

As scholars have emphasized, there is a need for analyzing the combination of resource specific property arrangements and government strategies with respect to natural resources. The literature calls this combination of property rights and policy intervention institutional resource regimes. Only on the basis of such an analysis is it possible to concentrate on the possibilities for public intervention in property rights, which adds the political steering dimension to the property rights approach. A strength of this perspective is that institutions, specifically property arrangements, are not just treated as frameworks within which actions are carried out, but also identified as the result and integral part of the political process.

The focus on the interaction of policy with property rights is important, because in developed societies guidance on heterogeneous, growing and increasingly rival use demands is required. In the case of homogenous demands for resource benefits discussed in much of the literature, scholars often found that it was possible to prevent the degradation of resources on the basis of voluntary cooperation, i.e. without state intervention. From a liberal perspective, this can be viewed as a very efficient strategy, of course. In the context of complex resource regimes, however, this solution is less likely to be feasible or effective. Here, the task of government intervention becomes to develop an integrated governance regime to avoid conflicts in property rights.[1]

This perspective on institutions as both the result and an important element of the process allows the analysis to avoid a fundamental weakness of many studies of resource regimes. The latter tend to focus on the analysis of regulative systems as they exist at one point in time. Lesser emphasis is generally placed on an analysis of processes of change. For the highest degree of policy relevance, however, i.e. to be able to avoid further degradation of resources, it is important to know when and under what conditions in the political process regimes can be changed and how this change can be accomplished and managed. The study of institutional resource regimes combining property rights arrangements with policy intervention allows the development of a foundation for such dynamic analyses.

This chapter, thus, focuses on the question of how governance strategies and property arrangements can come together to influence the sustainability of environmental resources. In pursuit of its objectives, the chapter applies two fundamental arguments. First, the analysis starts from the argument that from an economic or environmental perspective almost every policy intervention can be interpreted as a change in property rights, or, in fact, government intervention in property rights. Second, the analysis argues that traditional categorization of property regimes as private property, common property, open-access, and state-ownership can be transformed into two continuous variables, capturing the extent of collective action problems among the appropriators from a resource and the degree of state intervention in the property rights of those appropriators. On the basis of these arguments, the chapter then discusses to which extent government intervention in private property rights is environmentally desirable.

5.1 Policy as a change in property rights

From an *economic* and *environmental* perspective, almost every policy can be interpreted as a change in property rights. An institutional resource regime, thus, is determined at any point in time by how policies structure, i.e. create and influence, property arrangements. Policies shape property rights by intervening in specific parts of the bundle of rights held by the property owner. A per liter water charge imposed on withdrawal of water from the river by owners of real estate on the river banks, for instance,

[1] Given the distinction between de facto and de jure property rights discussed in chapter 2, the task of government may well be to avoid conflicts between de facto property rights rather than de jure property rights, at least in the short term.

establishes a new property situation. If the real estate owners were not charged for the water before, they either de facto "owned the right to free water withdrawal" or rights to the water were not defined. In the latter case, the water resource was an open-access resource of which the appropriators took advantage. After the policy change, the definition of property rights is clear. The real estate owners now own the right to obtain a given quantity of water for a given price. In fact, they can go to court and claim that right, if they are being charged more. Alternatively, the government can take them to court, if they refuse to pay the charge. For purposes of environmental management, the interpretation of the situation before the policy change does not matter. Real estate owners had free access to the water, and therefore no incentives to save water. For legal purposes, different interpretations of the situation before the policy change might matter, to the extent that the difference influences the possibilities of government to change the rights. After all, creating property rights in a previous open-access situation is less interventionist from a legal and political perspective than the appropriation of property rights previously held by private owners.[2]

Different degrees of policy intervention in private property rights, thus, may require different actions by government, and therefore different levels of political will and capacity depending on the constitutional framework. Sometimes governments can simply change specific property rights through simple policies, such as the implementation of water charges, for instance. In other cases, government might have to first pursue a constitutional change to be allowed a certain intervention in private property, or have to compensate the property owners. The extreme cases of government intervention in private property are, of course, expropriation and nationalization of private property. Because of their drastic nature and conflict with the constitutions of most democracies, these strategies frequently are not chosen. For most environmental purposes, however, they are also not necessary, as governments can achieve substantial changes in environmental management through simple policy modifications of property rights. As this theoretical framework focuses on environmental management rather than legal questions, the legal perspective of government intervention in property rights will not be pursued here. Rather, interpreting policy changes as changes in specific economic property rights, the analysis will explore the environmentally desirable extent of government intervention in those property rights.

[2] The counter argument is, however, that customary rights often can also be claimed in court.

5.2 The environmental desirability of intervention

The following discussion will show that the environmental desirability of government intervention in property rights depends on three specific factors, that nature of which is determined by the particular combination of the socio-economic context of the resource and its ecological characteristics. The three factors are the number of appropriators from the resource and the extent of collective action problems among them, the relationship between the economic and environmental values of the resource, and the level of government capacity and commitment to environmental objectives. The role of the relationship between economic and environmental values of the resource, the *e-e gap*, has already been discussed in chapter 4 and will be recapitulated only briefly in the following discussion. The other factors, however, will be laid out in some detail, before the analysis draws the different factors together to provide a comprehensive image of the environmental desirability of government intervention in property rights.

From property regimes to the 'Size of CAP'

As a first step in this analysis, the traditional categorization of property regimes as private property, common property, and open-access (Feeny, Berkes, McKay, and Acheson, 1992) is transformed into a measure of collective action problems. Rather than thinking of these property arrangements as categorically different, this analysis perceives them as stages on a continuum. After all, both private and common property regimes are essentially special cases of private property arrangements, with the difference that in the latter case the number of appropriators is greater than one and therefore the collective action problems and costs of governance are likely to be bigger.[3] Furthermore, the reasons for resource degradation of private property, i.e. 'lack of incentives to fight against negative externalities' can also be applied to common property.

The differences between private and common property regimes on the one side and open-access on the other are bigger. In the legal sense, open-access means that no property rights have been defined and that therefore nobody holds the rights to the given resource. Environmentally, however, the expected consequences of an open-access situation are similar to a common property regime that has failed to achieve a sustainable cooperative solution. Alternatively, even without the definition of legal rights in the Western sense or even tenure in a more traditional sense, appropriators from an open-access resource might achieve sustainable management of a resource similar to a successful common property regime.

[3] Eggertson (1996, 1998), by the way, points out that the costs of governance of private property rights are not to be underestimated due to the high costs of exclusion.

Environmentally, then, the difference between a common property regime and an open-access situation is that the former is more likely to succeed in the sustainable management of a resource, while the latter is more likely to result in its overexploitation. This dynamic can be expected because in common property regimes the group of appropriators would tend to be better defined, and therefore communication and cooperation are more easily established. In addition, outsiders can more easily be prevented from access to the resource. In other words, common property regimes generally have less collective action problems, due, for instance, to lower transaction costs. Fundamentally, however, common property regimes can be perceived as cases in which appropriators have developed successful institutions of governance, while in open-access situations these institutions have broken down or have never been created. In terms of a continuum, the transition from common property regimes to open-access corresponds to an increasing breakdown of institutions of governance among appropriators from resources.

The potential sustainability of environmental management in private property situations, again ranging from individual ownership to open-access situations, is, thus, a function of the collective action problems among the appropriators, in shorthand the *size of CAP*. Governance breaks down and common property turns into open-access when these collective action problems become too big to be solved. As Ostrom and others have shown, the *size of CAP* is determined by the characteristics of the appropriators from the resource, such as the number of decision-makers, the heterogeneity of the group in terms of capabilities, cultural and social values and preferences, information and beliefs, effective leadership, and the internal structure of the actors (Feeny, 1998; Keohane and Ostrom, 1995; Schlager and Blomquist, 1998). Furthermore, characteristics of the resource and socio-economic context, such as rivalry/subtractability of the benefits from the resource, its density and predictability, geophysical structures, in particular natural barriers to entry by outsiders, existing institutional arrangements, and exogenous determinants such as the technologies of extraction, monitoring and enforcement matter (Dasgupta and Mäler, 1995; Ostrom, 1990). For a mobile resource such a migrating deep-sea fish populations, for instance, collective action problems will, ceteris paribus, be bigger than for a non-mobile resource.

The *size of CAP* thus needs to be viewed as an index summarizing these variables and has to be determined in the context of a given natural resource. The analysis adopts this construct in the interest of simplification and illustration of the argument to highlight the range and type of factors influencing governance and exclusion costs. Based on the *size of CAP*, it is possible to align the three 'property regimes' along a continuum from small collective action problems with high probabilities of successful cooperation to large collective action problems with low probabilities of successful cooperation. The smaller the *size of CAP*, the more economically efficient the property arrangement.

From state ownership to government intervention

As a second step in the development of the argument, the analysis abandons the traditional fourth category of property regimes, state ownership, as well. 'State ownership' insinuates that an entire resource is owned by the state in contrast to resources held under private property or common property regimes in which the state is not present at all. Neither case is common in real life. If property rights are conceived as bundles of rights defined with respect to different attributes of a resource, it becomes clear that individuals rarely own rights to all attributes of a resource. This is partly the case because defining and enforcing property rights to all attributes would be prohibitively costly, partly because governments are unwilling to relinquish rights to certain attributes. For example, governments generally hold the rights to the airspace above 'private' land, or do not give citizens the right to use their 'private' ponds to dump toxic waste. In almost all cases of private property ownership the government retains some rights, precluding the owners of the remaining rights from a certain action or forcing them to execute another.

In common practice, government 'rights' often take the form of policies or regulations, but from an economic/environmental perspective, they are, in fact, rights to attributes of goods held by government. In the case of what traditionally is identified as state ownership of natural resources, the government just holds more rights and, most importantly, rights to more visible attributes of the resource. In consequence, the argument suggests transforming the category of state ownership into a continuum of government intervention, where a higher level of government intervention means that the government holds rights to more attributes of a resource. A low degree of government intervention, in contrast, refers to what traditionally is called private property, common property, or open-access regimes.

The degree of government intervention, then, forms a second continuum. It is fundamentally different from the *size of CAP*, since it captures the extent to which attributes are held by a formal authority. Any property arrangement with respect to a given natural resource is defined by the level of collective action problems among the appropriators *and* the degree of government intervention in the rights of these appropriators. Indeed, governments intervene to a similar degree at any *size of CAP*, or - if described in the traditional categories - in 'private property,' 'common property,' or 'open-access.'

State capacity and commitment to sustainability: the c-c level

The third step in the development of the argument introduces government capacity and commitment to sustainability (in other words, government's will and skill) as an additional determinant of the environmental implications of property arrangements, represented by the *c-c level.* The *c-c level* of a government determines the environmental implications of government intervention. Government intervention is often considered as an alternative to private ownership, in cases in which the maximization of private economic gain leads to the undersupply of an environmental good or the oversupply of an environmental bad. The capacity and commitment of governments to protect their own property rights and to use environmental resources in a sustainable manner determine the chances that government intervention does indeed lead to an environmentally superior outcome in such a situation. Government intervention has a high potential for environmental stewardship *only* if the respective government has high levels of capacity *and* commitment. The lower the *c-c level* of the state, ceteris paribus, the fewer resources should be subject to government intervention and the more should be held in private property arrangements.

Again, the *c-c level* is a construct summarizing the two independent variables, capacity and commitment. Please note that their effects are combined in the *c-c level* purely for illustrative purposes, as these variables do not necessarily move together. In fact, future research needs to examine possible interactions between the two variables. An interactive model might be able to show that the impact of each of the variables on environmental outcomes depends on the level of the other. In other words, the effect of state commitment might increase systematically as a function of state capacity and vice versa. Furthermore, both of these concepts should be viewed as continua rather than as dichotomies.

A high level of capacity, i.e. skill, captures the ability of government to achieve its desired policy outcomes. A high level of commitment, i.e. will, refers to the mission of the government in terms of maximizing public environmental welfare rather than private gain. In the context of this development of the argument, states with high *c-c levels* are those which have *both* the capacity *and* the commitment to protect their property rights to attributes of natural resources and to use these resources in an environmentally sustainable manner. States with low *c-c levels*, in contrast, are those where the government *either* lacks the capacity to protect government rights to natural resources, *or* pursues the maximization of private economic gain rather than public environmental welfare *or* both.

Thus, commitment is a necessary but not sufficient condition for an environmentally desirable impact of government intervention. The same applies to capacity, with the caveat that capacity without commitment is likely to have worse environmental impacts than commitment without capacity, as discussed below. The relationship between high and low *c-c levels* is, of course, not a dichotomous one, but continuous. Most governments in the world do not fall on either end of the spectrum but somewhere in-between.

In an excellent analysis of the role of the state in the creation and enforcement of property rights for economic growth, Kobler (2000) differentiates between two similar aspects of property rights. Kobler argues that both the 'Macht' (power) and the 'Bindung' (commitment) of the state are necessary requirements for the state to be able and willing to efficiently create and enforce property rights. Similar to the argument made in this book, Kobler points out that power without commitment holds the threat of an abuse of state power, i.e. the pursuit of its own material or political interests rather than those of the population. The crucial difference between the two arguments, however, is that Kobler limits himself to the question of the efficiency of property rights institutions.[4] He does not consider the role of government in the utilization of its own property rights in the pursuit of public goods such as environmental quality. In consequence, Kobler's concept of 'Bindung' is limited to the question of the control of government by functioning political structures and processes. This chapter's argument, in contrast, includes an environmental objective in its concept of commitment. Generally, of course, the environmental objective should originate with the public, which the state represents, and thus reflect the same aspects of control and responsiveness of the state that Kobler highlights.

Government *capacity* is a necessary condition for a high potential of government intervention for environmental stewardship because only a capable government will be able to enforce rights and regulations and be able to protect or provide an environmental good, which is threatened by private interests. In the absence of government capacity to enforce its own property rights, 'state-owned' property automatically becomes *de facto* an 'open-access resource,' because of the concurrent lack of any private incentives to protect the property rights (Deacon, 1994). As shown above, the nationalization of natural resources in developing countries has frequently had devastating environmental consequences (Adger and Luttrell, 2000; Ostrom, 1990). Thus, compared to common property situations, which might face collective action problems but often can draw on established systems of authority in the community and a supportive legal environment,

[4] Thus, he is a welcome reminder that these two aspects are also important with respect to the role of government in providing what was termed 'assurance' in the previous chapter.

state-ownership by a government lacking will and skill appears less desirable (Chakraborty, 2001).

General government *capacity* is a function of the availability of resources to government, such as material and human resources, and the efficiency of their utilization (Organski and Kugler, 1980; Arbetman and Kugler, 1997). Government capacity is, to a large extent, determined by its perceived legitimacy and authority and its corresponding support in the population and among powerful (military or economic) elites (Jackman, 1996). If a government lacks support within the country, its activities become costly and, ultimately, weaker. Press highlights this aspect when he defines (local) environmental policy capacity as a community's rather than just the government's "ability to engage in collective action that secures environmental public goods and services" (Press, 1998: 37). Accordingly, a community's social capital and environmental attitudes and behavior are important factors in his approach, besides political leadership and commitment, and administrative and economic resources. If we want to explore the question of capacity from the governmental rather than the community's role, then, we can recognize the extent to which community characteristics will influence this capacity.

Specific government capacity, i.e. government skill with respect to its involvement in a particular issue or issue area, furthermore, is a function of the validity of the policy strategy chosen, its implementation, and its integration with related policies (Bressers and O'Toole, 1998). Government weakness with respect to these aspects does not only have a direct negative impact on sustainable use. Rather, such a weakness also has the implication that the mix between private property rights as such and government intervention should be different.

In a different approach to the definition of the determinants of the state's ability to create and enforce property rights, Kobler (2000) uses an economic argument. He claims that this ability of the state is a negative function of the opportunity costs faced by the state and a positive function of the opportunity costs faced by the population when fighting against the state's creation and enforcement of property rights. The state's opportunity costs, in turn, are a function of the form and historical development of the political institutions, the homogeneity of individual preferences, the distribution of income and wealth, the geographic and demographic make up of the country and the quality of the infrastructure which influence the transaction costs associated with the creation and enforcement of property rights.

Government *commitment*, in the context of this chapter, identifies whether the government is likely to maximize public environmental welfare or private economic benefit. If a government is interested in deriving the maximum economic benefits from environmental resources under its control rather than protecting the public good of environmental quality, government intervention is similar in its environmental implications to unregulated private property arrangements. The predatory governments with no interest in environmental objectives, but rather in placating allies and maintaining patron-client relationships, in Southeast Asia are an excellent example for such a case (Dauvergne, 1997). For a government that is weak in terms of commitment to sustainability, government intervention will be less environmentally desirable than for a government that is strong in this respect.

Different degrees and causes of weakness of governments in terms of *commitment* can be imagined. On the one side, governments might give in to pressure from special interests and therefore might not be sufficiently sensitive to the question of environmental quality. On the other side, governments might pursue their own financial gain through the unsustainable exploitation of 'state-owned' natural resources. From those governments, it is sometimes not a big step to governments appropriating previously private property for their own gain. Kobler (2000) uses an economic definition of commitment. He argues that states will be prevented from abusing their power if the costs of doing so outweigh the benefits. In terms of costs, Kobler refers to Hirschman's (1970) 'exit,' 'voice,' and 'loyalty.' Thus, the population can create costs for a state abusing its power through 'exit' (emigration) and 'voice' (attempts at inducing a change in the state's behavior).

In some developing countries, a lack of government *commitment* has been reflected in an intentional failure of governments to recognize traditional (tribal or communal) rights of ownership to natural resources. Governments have frequently appropriated such resources, in order to increase governmental or personal revenues or to sustain political support by giving these resources to political and economic allies. A lack of commitment based on such a preference structure of the governing elites will likely carry the worst consequences for environmental stewardship as it does not only affect the environmental fate of 'state-owned' resources but of all resources. Predatory behavior by government has two consequences. Since the new 'owners' of the resource are generally aware of the illegitimacy of their rights and the lack of accountability of the government, they have an incentive to deplete the resource as fast as possible.[5] After all, they know that they might fall out of favor with the government themselves, or fear the

[5] In practice, the reaction of the traditional owners of resources to government appropriation of property rights is often environmentally less dramatic than that of the illegitimate new owners, since the former frequently depend on the existence of the resource for their livelihood. In economic terms, their choice set as rational decision-makers is more constrained than that of the 'new owners.'

overthrow of the government by a disenchanted public. Likewise, the traditional owners of resources have similar incentives to deplete them as fast as possible, if they witness such predatory behavior by government, since they have to fear that they will be the next ones to lose their rights. The higher the chances for resources to be appropriated by governments, the higher the discount rate owners will apply to future revenues from resources (Deacon, 1994).

Given the constraints *capacity* and *commitment* impose on the potential of government intervention for environmental stewardship, it becomes obvious that a high degree of government intervention in private property arrangements is not necessarily environmentally superior to a low degree of government intervention. Indeed, only in situations in which a capable government pursues the public good of environmental quality has government intervention the potential to improve on environmental outcomes from private ownership. Ceteris paribus, a high *c-c level* of the government increases the environmental desirability of government intervention in private property arrangements.

Recapitulation: the e-e gap

The fourth variable influencing the environmental implications of property arrangements and therefore the environmental desirability of government intervention is the *e-e* gap, which was introduced in chapter 4. Recall that the *e-e* gap is the difference between the 'economic value' of a resource and its 'environmental value,' with 'economic value' referring to *the maximum economic value to be obtained from any use of the resource,* while 'environmental value' refers to *the economic value that can be obtained from the environmentally most desirable use of the resource.*

To recapitulate, the *e-e gap* is important for the environmental desirability of property arrangements, because it determines the implications of the maximization of expected utility for environmental stewardship. If the economic and environmental values of a resource are close, the decision-maker's maximization of expected (economic) utility is more likely to imply a maximization of environmental stewardship, and therefore the most efficient property regime in economic terms will also be the most desirable one environmentally. If, however, the difference between the two values is large, the maximization of expected (economic) utility is likely to result in the maximization of environmental degradation, and therefore the economically most desirable property regime will not be the environmentally most desirable one.

As an example, consider the following. For the owner of a small fishing ground next to the house in the middle of nowhere, the *e-e gap* is small. The owner is likely to use the fishing pond for fishing and has incentives to fish only as much fish out of the pond that a population capable of renewing itself remains. If the fishing pond is in the middle of Manhattan, however, and billions of dollars could be made by draining it and building office and apartment buildings, the *e-e gap* is huge, since the environmental and economic values of the fishing pond differ vastly. In the first case, private property with a low level of collective action problems is the economically most efficient arrangement and also allows the highest degree of environmental stewardship. In the second case, an economically efficient private property arrangement is not appropriate to protect the environmental value of the fishing pond, but rather most likely to lead to its degradation. Under such circumstances, government intervention in private property rights would be environmentally desirable.

In other words, the *e-e gap* determines if the economically most efficient property arrangement is the environmentally most desirable one. With the aid of the *e-e gap,* one can show that a small *size of CAP* only implies a greater potential for environmental stewardship in certain cases, specifically in cases in which the *e-e gap* is small. In cases, in which the *e-e gap* is large, a small *size of CAP* implies a greater potential for an economically efficient environmental degradation of the resource. Here, government needs to intervene on behalf of sustainability.

Pulling it all together

These four variables, the level of collective action problems among appropriators, the degree of government intervention, government capacity and commitment to sustainability, and the relationship between the environmental and economic values of a given resource determine the environmental implications of any property arrangement. Since the analysis is primarily interested in potential policy implications, the degree of government intervention will be treated as the dependent variable in the following discussion. Based on the above analysis, the environmentally desirable degree of government intervention is a function of the *size of CAP*, the *c-c level* of the government, and the *e-e gap*, in the context of a given resource.

The arguments can be summarized in the following hypotheses:

1. The higher the *c-c level* of the state, ceteris paribus, the higher the environmentally desirable degree of government intervention.[6]

[6] Future research needs to look at the relationship between c-c levels and environmental state intervention to determine if strategic behavior on the part of potential environmental 'villains' exists. Strategic polluters, for instance, could anticipate a response by a state with high *c-c levels* and therefore reduce their environmentally polluting activities. In other words, while the environmental desirability of government intervention in theory is

2. The larger the *e-e gap*, ceteris paribus, the higher the environmentally desirable degree of government intervention.
3. Given a small *e-e gap*, the larger the *size of CAP*, the higher the environmentally desirable degree of government intervention.
4. Given a large *e-e gap*, the smaller the *size of CAP*, the higher the environmentally desirable degree of government intervention.

Figures 1 and 2 combine these hypotheses, depicting the interaction between the *c-c level* of the state, the *size of CAP*, and the *e-e gap* and their implications for the environmentally desirable level of government intervention. These figures illustrate under which conditions a larger share of the attributes of a given resource needs to be protected by government intervention for sustainability. The separation of the argument into figures for large *e-e gaps* (Figure 1) and small *e-e gaps* (Figure 2) allows the visual illustration of the hypotheses in a three dimensional space. The reader is asked to keep in mind, however, that this simplification hides some of the complexity of the situation, since the *e-e gap* is not dichotomous but continuous in reality.

Figure 1 demonstrates that for a large *e-e gap*, the highest degree of environmental desirability of government intervention applies to cases in which a high *c-c level* is combined with a small *size of CAP*. In other words, if the economic and environmental values of a resource diverge substantially, a high degree of government intervention in private property arrangements is most desirable if the government itself is capable and committed to sustainability, and if a low degree of collective action problems among the appropriators (such as individual ownership) suggests that the economically efficient environmental degradation of the resource is likely. This is the case with respect to Central Park in New York, for instance. If a private individual was able to acquire a real estate plot in Central Park, it is very likely that this individual would try to reap the probably very large economic benefit of developing it for residential use, for instance. Here, government intervention is necessary to protect the environment.

The lowest degree of environmental desirability of government intervention applies to the opposite case. If a low *c-c level* is combined with a large *size of CAP*, government intervention will not be able to significantly increase the potential for environmental stewardship, and indeed might lower it. In other words, in case of a substantial divergence between the economic and environmental values of a resource, a weak and/or environmentally neglectful or predatory government will probably not

higher the higher the *c-c level*, in practice high c-c levels might be associated with lower levels of actual intervention. This would suggest a non-linear relationship.

improve the potential for environmental stewardship much compared to a group of appropriators with large collective action problems. As an example, consider the case of the Malaysian rainforest. Here, the unsustainable forest management by villagers was much surpassed in the destructiveness of harvesting by a predatory government, which valued the short-term benefit of payments for timber much more than the long-term ecological benefits of the rainforest.

Thus, when deciding the environmentally desirable degree of government intervention in the context of a large *e-e gap* it is important to keep in mind that a small *size of CAP* is likely to lead to the most efficient environmental degradation. In contrast, a large *size of CAP* has the potential to inhibit the most efficient exploitation of this resource, thus, requiring less government intervention to protect the resource. Furthermore, a state with a high *c-c level* is more likely to pursue and achieve environmentally desirable outcomes than a state with a low one. Cases for which a low *c-c level* is associated with a small *size of CAP*, or in which a high *c-c level* is associated with large *size of CAP* fall somewhere in the middle between the two extremes discussed above.

In Figure 2, the presence of a small *e-e gap* changes the situation dramatically. Here, the highest degree of environmentally desirable government intervention results from the presence of a high *c-c level* and a large *size of CAP*, because the latter leads in this case to the largest waste of natural resources. In other words, if the economic benefit of a resource is closely associated with its environmental health, a capable and committed state should help a group of appropriators with large collective action problems (large enough that the group cannot solve them itself) overcome these. A typical situation would be off-shore fishing grounds, for instance, always keeping in mind the necessary will and skill of the state, however.

Since a small *e-e gap* means that the economic efficiency resulting from a small *size of CAP* translates into environmental 'efficiency,' the environmentally desirable level of government intervention in the case of a small *size of CAP* is lower. Again, in this case of a close alignment between environmental and economic efficiency, government intervention is only necessary to help solve collective action problems so that efficient outcomes can be achieved. If collective action problems are small (because the fishing ground is only utilized by one individual or a few individuals, for instance), government intervention is less important.

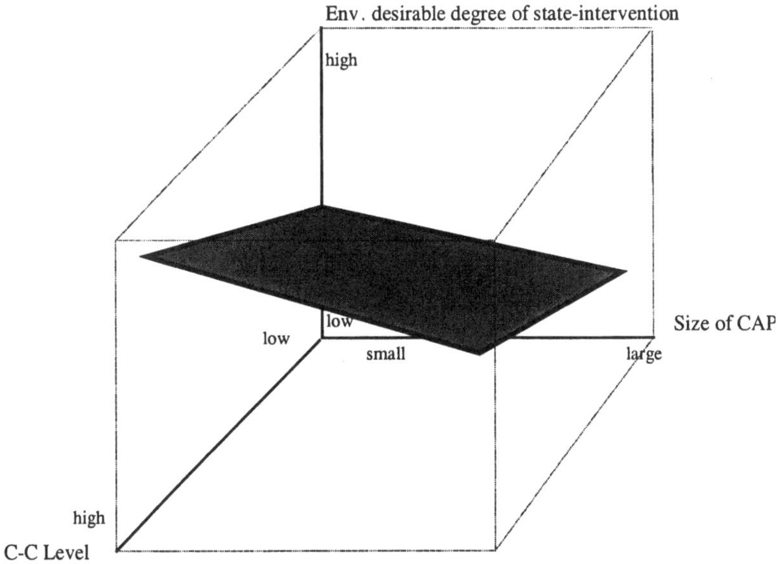

Figure 1. Large E-E Gap

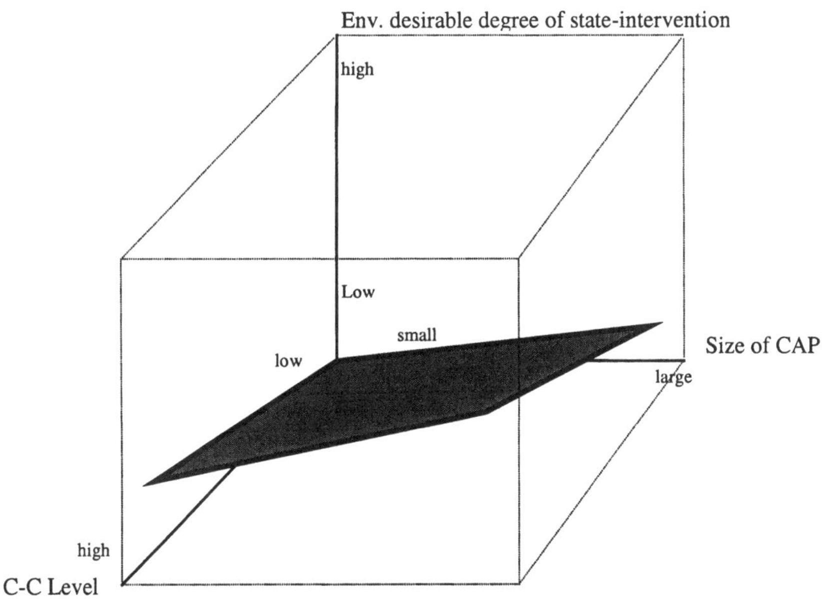

Figure 2. Small E-E Gap

The lowest degree of environmental desirability of government intervention therefore applies to cases in which a small *size of CAP* is matched with a low *c-c level.* Not only is government intervention not very necessary in this case, but it is extremely undesirable. The appropriators are likely to arrive at an environmentally and economically efficient outcome themselves. If government took over, however, its lack of capacity and commitment to environmental stewardship would likely induce a less sustainable outcome. Government intervention in natural resources previously owned by tribes or villages in many developing countries has proven exactly this point (Adger and Luttrell, 2000; Dauvergne, 1997; Ostrom, 1990). Chakraborty (2001), for instance shows that common property forests fare better than state-owned forests in Nepal.

As expressed by the two figures, higher *c-c levels* are, ceteris paribus, associated with higher degrees of environmental benefits of government intervention. Putting it more controversially, the figure highlights the limits of the environmental desirability of government intervention if a government is weak in terms of capacity or commitment. Looking at government capacity and practices in countries around the globe, these conditions would apply to a large share of global resources.

The influence of the *e-e* gap is more difficult to gauge, because of the necessary comparison across the two figures. This comparison shows, however, support for the claim that a large *e-e gap*, ceteris paribus, is associated with a higher level of desirability of government intervention. Thus, the *e-e gap* highlights the limitations of private property rights as a means to raise the potential for environmental stewardship, just as the *c-c level* identifies the limitations of government intervention.

Hypotheses 3 and 4 capture one of the most intriguing aspects of this analysis. The figures illustrate that the degree of environmental desirability of government intervention is larger for a small *size of CAP* if the *e-e gap* is large, but may be smaller for a small *size of CAP* if the *e-e gap* is small. This finding is in stark contrast to the agreement in the literature that open-access resources generally fare the worst. The above analysis suggests that this is not necessarily the case. A small number of appropriators from a resource can be associated with worse degradation if the increase in efficiency in the maximization of the economic value of the resource is associated with an increase in the efficiency of environmental degradation. The neglect of this dynamic in the literature results from the fact that the majority of resources that are usually considered are resources under some form of environmental use, where the difference is in degree of sustainability of that use (such as land used for agriculture, or forests used for timber extraction). Again, the analysis highlights that a small number of appropriators from a resource will only necessarily be associated with higher environmental benefits, if the environmental costs and benefits of actions are fully internalized, which would render the *e-e gap* small.

5.3 Implications

This discussion, hopefully, has demonstrated that the environmental implications of property arrangements need to be studied in context. When looked at from the perspective of the environmental desirability of state-intervention, this argument has deduced three factors that need to be considered. One factor fundamentally influencing the desirability of state-intervention is the *e-e gap*. It is a surprisingly basic factor in the determination of the environmental implications of property arrangements, and yet (or maybe therefore) is rarely being explicitly analyzed and discussed. If the *e-e gap* is small, i.e. the correlation between the environmental and economic values of a resource is high, state-intervention will, ceteris paribus, be less environmentally desirable, especially the smaller the degree of collective action problems. The second factor is the *c-c level* of government. The higher the levels of capacity and environmental commitment of a government and the larger the e-e gap, the higher the environmental desirability of state-intervention. Thirdly, the level of collective action problems needs to be considered. The smaller the *size of CAP* given a small *e-e gap*, the lower the environmental desirability of state-intervention.

The implications of this analysis are very informative with respect to the property rights debate. They show how the various arguments and evidence from previous debates and research can be integrated into one cohesive picture. Thus, the analysis supports findings in the resource economics literature that show that there is neither a theoretical nor an empirical reason for believing that one property regime is generally environmentally superior to another. As Devlin and Grafton (1998) state there is no 'best' regime from an environmental perspective.

Furthermore, by transforming the traditional categorization of property regimes into two continuous variables, the *size of CAP*, and the degree of government intervention, the analysis can focus on key determinants of the environmental implications of different property arrangements: the *c-c level*, and the *e-e gap*. In addition, the argument illustrates the dynamic nature of the environmental implications of property arrangements, allowing scholars to capture changes in these implications as a function of changes in specific underlying conditions over time. Thus, the project allows taking a step in the desired direction as identified by Schlager and Ostrom:

> Instead of blind faith in private ownership, common-property institutions, or government intervention, scholars need a better understanding of: (1)

> the conditions that enhance or detract from the emergence of more efficient property-rights regimes related to diverse resources, (2) the stability or instability of these systems when challenged by various types of exogenous or endogenous changes, and (3) the costs of enforcing regulations that are not agreed upon by those involved (1992: 260).

The argument also makes explicit the potential and limitations of various property rights based strategies for improving environmental stewardship. First, the analysis highlights that for government intervention to be environmentally desirable, governments have to be both committed to environmental goals and capable to effectively pursue their policy objectives. Given political conditions around the world, this requirement may place severe limitations on the desirability of government intervention on behalf of the environment. Many governments still do not necessarily regard environmental protection as an important public good and policy objective. This is partly the case, because they believe that other policies, especially economic policies, are more important for the welfare of their populations, but often because the governing elites are primarily interested in their own (financial) welfare. The Indonesian government's practices in distributing timber licenses are a case in point. Likewise, many governments lack the capabilities to make state-intervention on behalf of the environment a success. This analysis, therefore, shows that critics of government intervention are correct in highlighting its limitations, and there is ample evidence from the nationalization and subsequent degradation of environmental resources, especially in developing countries, to support their argument. Recognizing the impact of capacity and commitment in determining the potential and limitations of state intervention highlights the importance of further research on the determinants of these two variables.

At the same time, the analysis underlines that these conclusions cannot easily be generalized across cases or countries and that government intervention can indeed be environmentally desirable. If governments are both capable and committed, government intervention can prevent or at least reduce environmental damage by unregulated private ownership. Importantly, one finds the closest match for such conditions especially in those countries where the costs of government intervention and the benefits of private ownership are proclaimed most loudly: the industrialized countries, especially the US. Here the critics of private property rights are correct when arguing that government intervention can have the potential to increase the public good of environmental welfare - under certain conditions!

Secondly, the analysis highlights the potential and limitations of privatization and improvements in the assurance of property rights for environmental stewardship. In other words, the argument illustrates under which circumstances an increase in individual control over natural resources, as advocated by proponents of privatization and secure property rights will make government intervention more or less environmentally desirable. If the *e-e gap* is small, the advocates of private property rights are correct, and therefore one will find examples cited in support of their arguments to be characterized by the respective conditions: small *e-e gaps* (and/or low *c-c levels*). As the analysis has shown, a small number of appropriators from a resource are unequivocally preferable in the case of a small *e-e gap*. Given a significant divergence between economic and environmental values of a resource, however, the number of appropriators from the resource has ambiguous implications for environmental stewardship. In the latter case, governments would probably fare better in setting up institutional support for governance problems (which can draw on the capacity of the community if the central government is lacking capacity) than in 'privatizing' the resource. Alternatively, for governments with sufficient capacity and commitment, privatization could be accompanied by the retaining of specific rights by the state, i.e. government intervention ensuring sustainable resource use.

Even more, the analysis suggests that rather than viewing a resource as existing under one particular property regime, it is important to differentiate between the different attributes of the resource and identify the environmentally desirable owner or owners for those. Thus, some parts of the bundle of rights associated with a resource might be held by the state, others by a group of owners, and others again by individual owners (see also Lueck, 1995). Princen (1998) follows a similar notion, when he suggests that generalizations on the basis of property regimes are not necessarily convincing due to the large difference among resources and resource characteristics. He, therefore, argues for the delineation of specific rights *and responsibilities* with respect to a given resource. The specific distribution of rights needs to be analyzed on the basis of the factors identified in this discussion; the *c-c level*, the *e-e gap*, and *size of CAP*.

In sum, general statements can be derived which allow interesting insights into the desirable level of state-intervention for environmental stewardship. Most fundamentally, the advantage of these general statements is that any bias associated with case selection can be avoided. The logic can be applied to any resource independent of type, size, and form of ownership. This has the additional advantage that research on the environmental implications of property arrangements can utilize a larger n and control for the same fundamental influences across the sample. The efforts of Ostrom's group to determine the conditions under which common property management is likely to be successful can be enhanced, for instance, through the utilization of this general approach which allows to control for the various influences such as the *e-e gap*. In addition, the logic illustrates the dynamic nature of the environmental implications of property arrangements, allowing scholars to capture changes in these implications as a function of changes in specific underlying conditions over time.

The next major step in the pursuit of this research would need to be, of course, the operationalization of the argument. The empirical application is associated with some difficulties, but it is possible. To begin with, measures of government capacity have been developed and applied, most prominently by Kugler and Arbetman (1996). An approximation of government commitment could be estimated, for instance, as a function of either the government budget spent on environmental protection or the willingness of governments to become a party to international environmental agreements. While none of these measures are perfect, they would be a start. The *size of CAP* could be estimated at the current stage. Finally, the *e-e gap* could be gauged on the basis of the expected value of market returns on the maximization of the economic value of the resource, and the expected value of market returns on the environmentally most desirable use of the resource, given expert judgment of what the latter would be. As this brief discussion of matters of operationalization shows any empirical analysis of the argument would need to take place in a temporal context. The values of the *c-c level* and the *e-e gap* in particular are constantly changing. *C-C levels* can change with every change in government if not more often. The *e-e gap* will change with changes in market prices. While these effects do not make the empirical application of the argument any easier, they do capture the true dynamic nature of the problem.

5.4 Conclusion

This chapter has developed a general approach to assessing the environmentally desirable degree of government intervention in 'private' property rights. It has demonstrated that the extent of desirable government intervention is a function of the capacity and commitment to sustainability of the government (the *c-c level*), the relation between the 'economic' and the 'environmental' values of a resource (the *e-e gap*), and the extent of collective action problems among the appropriators of the resource (the *size of CAP*). As these factors change, the need for government intervention in private property arrangements on behalf of sustainability changes. Thus, any analysis of the environmental implications of a given property arrangement needs to consider the three factors and take into account the respective temporal dynamics.

In general, government intervention on behalf of the environment is the more desirable the higher government capacity and commitment to sustainability and the more economic and environmental resource values diverge. Moreover, if economic and environmental resource values do not differ much, government intervention is most desirable in the presence of large collective action problems among the appropriators from the resource. Here, government can help achieve an economically and environmentally efficient outcome. If economic and environmental resource values differ substantially, however, government intervention is even more desirable in the presence of small collective action problems. In this case, government intervention needs to avoid an economically efficient environmental degradation.

In sum, this analysis has numerous implications for previous and future research on the environmental implications of property rights. For research on the implications of security of property rights (Deacon, 1994, 1995) and of individual property rights (Gordon, 1954; Hardin, 1968), the analysis highlights the decisive function of the *e-e gap*, which can 'turn' a positive environmental impact into a negative one. For participants in the public vs. private property rights debate, the model shows that the choice is not one or the other. In addition, it avoids the ideological bend of the debate by generalizing the question of the environmental desirability of state-intervention beyond private property regimes, and formalizes the evaluation of the benefits of state-intervention. For the CPR school, the analysis allows the extension of inquiry beyond resources under joint management, and more importantly, formalizes the evaluation of policy options if joint management fails. In the latter case, three options to limit or prevent environmental degradation exist. First, the creation of conditions that the CPR school has identified as determinants of successful joint management could be attempted. Secondly, the state could be required to intervene on behalf of the environment, and thirdly privatization of the resource could be attempted. The approach presented here formalizes the evaluation of the impact of the latter two options.

Finally, the implications of this study reach beyond the topic at hand. The e-e gap, for instance, is important not just for the environmental desirability of state-intervention, but for all studies of the relationship between property rights and the environment. Likewise, the theoretical foundations of the model with respect to the state, i.e. the importance of government capacity and commitment, are likely to apply to a range of policy areas. The discussion already noted their importance with respect to assurance of property rights. But government capacity and commitment also matter in non-environmental policy arenas. Economic growth, for instance, would appear to be a primary candidate to compare government capacities and levels of commitment. Such an analysis could inform foreign aid policy as well as foreign direct investment decisions.

Chapter 6

Conclusions

6.1 The current stage: what did we learn?

The analyses in this book hopefully have demonstrated that by modifying the quality and structure of property rights, governments can improve the potential for environmental stewardship. In other words, by adjusting the institutional framework governments can induce state and non-state actors to improve environmental quality at a given level of development and for natural resources under any 'property regime.' Most importantly, the findings highlight the environmental impact of politics and thereby complement previous studies on the role of per capita income levels, but also on other political, social, and cognitive factors. Questions of governmental assurance of property rights, governmental intervention in the structure of property rights, and governmental capacity and commitment are shown to be important determinants of environmental stewardship at any given level of development.

How do these dynamics work? Improvements in the assurance of property rights improve the potential for environmental stewardship by changing the incentives decision-makers face when maximizing their benefits with respect to natural resources. By awarding property owners with long-term control over the revenues from a resource, the assurance of property rights induces them to make withdrawal and consumption decisions accordingly and improves the chances that they will keep the resource equilibrium at a sustainable level. Moreover, improvements in the assurance of property rights reduce transaction costs and increase the expected benefits from natural resources. Thereby, they induce rational decision-makers to define property rights to a greater extent and reduce the amount of resources for which the costs of defining and enforcing property rights outweigh the benefits. In consequence, a smaller amount of resources remains as environmentally-vulnerable open-access resources. Fundamentally, the assurance of property rights structures the choice set of owner or decision-maker with respect to natural resources under any type of property regime, generating an improvement in the potential for environmental stewardship across property regimes. Importantly, assurance of property rights is not a sufficient condition for environmental stewardship and can in cases of a large divergence between the economic and environmental values of a resource foster degradation. Thus, assurance of property rights should ideally be increased in conjunction with improvements in the internalization of environmental costs and benefits.

Governmental intervention in property rights, in turn, can improve the potential for environmental stewardship by protecting attributes of resources most threatened by the pursuit of private economic gain. In cases, in which economic gain is most closely aligned with environmental degradation, government intervention in incentive structures can foster improvements in the sustainability of resource management. Likewise, government can facilitate the overcoming of collective action problems if those hinder the achievement of environmentally and economically desirable outcomes, i.e. when economic gain from a resource is closely associated with its environmental health. Such intervention should not be confused with the turning of natural resources under common property into state-owned property, which in many cases has been found to lead to even worse environmental consequences. The analysis, thus, introduces a new perspective on the environmental desirability of governmental intervention in 'private' property rights and thereby contributes to the ongoing attempts in the discipline to move away from an undifferentiated foci on 'property regimes' to a more detailed evaluation of the environmental impact of specific property rights structures under particular economic, political, and ecological conditions. Highlighting the impact of the degree of convergence between the environmental and economic values of a natural resource, the level of government capacity and commitment, as well as the extent of collective action problems among appropriators, the analysis suggests a provisional ranking of the environmental desirability of different structures of property rights in different contexts.

Finally, the analyses have found that governmental capacity and commitment are important influences on environmental stewardship. Governments need to be both capable and committed to providing assurance of property rights on the one side, and to intervening in property arrangements in pursuit of environmental stewardship on the other. Given the variance in these conditions around the world, environmental NGOs as well as the international community as such may want to consider how to improve the worst cases.

These forms of governmental activities, i.e. the structuring of the institutional setting, are not sufficient conditions for environmental stewardship and sustainability, of course. Additional forms of government intervention are needed, most importantly the pursuit of measures fostering the internalization of environmental costs and benefits. As both the analyses on the environmental impact of assurance of property rights and on the environmental desirability of government intervention in the structure of private property rights have shown, the alignment of economic and environmental gains would greatly assist any institutional basis provided for the support of environmental stewardship in achieving its objective.

The policy advice resulting from these findings, then, is two-fold. First, with respect to assurance of property rights, governments could attempt to improve the environmental performance of actors by improving the assurance of property rights. Moreover, the analysis suggests that developed countries concerned about environmental problems in developing countries could help the latter in improving environmental conditions by investing in a strengthening of the legal system of the respective countries. This policy strategy is limited to countries lacking the capacity to assure property rights, of course, and cannot be used for countries in which the government or political elites choose not to assure property rights for ideological or personal reasons. Moreover, this strategy might not be politically, ethically, or even environmentally advisable if the improvement in the legal system merely strengthened the monopoly control of resources of politically unstable elites who pursue private economic interests. Likewise, the results indicate a need for the globalization of the assurance of property rights to ensure that differences in assurance between countries do not lead to an exporting rather than a reduction of environmental degradation. After all, local improvements in environmental quality that are associated with improvements in assurance and/or per capita GDP could, in the worst case, simply be the results of the moving of the most polluting industries to other countries or regions.

Secondly, with respect to government intervention in property arrangements, the analysis suggests that governments and other interested parties conduct a differentiated analysis of the need for intervention on the basis of the combination of present and likely future relationship between economic and environmental resource values and the extent of collective action problems among appropriators from the resource. Outside observers, in particular environmentally concerned individuals and NGOs, will also want to take governmental capacity and commitment into account when arguing for or against government intervention in such cases. In the case of weak capacity or commitment, they may want to consider strategies for improving this situation.

6.2 The next stage: where do we go from here?

As with most research projects, the analyses conducted here open up numerous new questions and thereby lead to the development of a new research agenda. First, the sources of assurance of property rights but also of governmental capacity and commitment need to be explored further. The present analyses already addressed a number of variables - albeit in terms of their relationship with environmental quality - which are likely to be of importance in answering that question. The choice of government regime, cultural determinants, income distributions, and even technological innovation all can potentially have an impact on the assurance of property rights, as well as capacity and commitment. While we tend to distrust authoritarian regimes in terms of their willingness to respect private property rights, an argument could be made that under certain conditions a democracy might not do well in achieving a high level of assurance or capacity either. Likewise, the interaction between income distributions and the assurance of property rights as well as governmental capacity needs to be considered. We have to ask what the implications of assurance are if access to legal recourse, for instance, is conditioned by wealth. Furthermore, the scale of the community is likely to influence the political or social level at which assurance is needed. Finally, the discussion of the determinants of assurance of property rights by government has highlighted the need to address the relationship between political stability, the capacity of governments, and environmental quality. Likewise, the determinants of the intervention of government in private property arrangements on behalf of the environment need to be further examined. Chapter 5 inquired into the determinants of the desirability of such behavior. It did not speak to the factors influencing whether the government will actually pursue such strategies.

Secondly, the analyses need to be applied to empirical assessments for a range of indicators of environmental quality. Most importantly, those environmental measures looking beyond local pollution and therefore invulnerable to biases resulting from the exporting of environmental degradation need to be considered. Such indicators include carbon dioxide emissions, energy and material consumption, and, ideally, indicators deriving from the ecological footprint analysis. In the pursuit of the necessary formalization of the arguments and especially an empirical operationalization, we face serious challenges, of course. For assurance of property rights, an adequate indicator is still lacking. Indicators of government capacity and commitment are also not as reliable as one would like them to be (see Appendix A). Moreover, the respective empirical assessments could not be situated at the aggregate level, but would require resource-specific data. After all, the degree of convergence between the economic and environmental values of a resource will vary greatly with the given socio-economic context. Moreover, the operationalization will depend on substantial advancements in the field of valuation of environmental resources. While a perfect measurement of the convergence between the environmental and economic values of a resource does not seem feasible at any point in the near and possibly more distant future - even two environmentalists are rarely able to agree on the environmental value or environmentally-desirable use of a natural resource - progress in this area is possible and necessary (and currently taking place).

Finally, future research needs to consider the implications of these analyses for international environmental politics and the management of the global commons. Given the prevalent lack of property rights to the global commons, the applicability of the results on the impacts of assurance of property rights and the environmental desirability of government intervention in property arrangements is not immediately obvious. However, the creation of property rights increasingly is becoming a possibility in global environmental politics, for instance in the form of exclusive economic zones and quotas for fishing, or emissions trading under the Kyoto Protocol. Thus, clarity of definition of property rights and the provision of assurance may help with respect to fishing areas or quota, where economic benefit and environmental health are likely to be closely associated. The same is the case for emission credits, since the latter are structured in a way that environmental and economic incentives follow the same direction. Assurance of property rights would probably help little with respect to tracts of rainforest, however, since the economic gain from cutting down the forest still outweighs the environmental benefit an individual could gain from its sustainable management by far. Similar to domestic contexts, then, assurance of property rights could not provide a sufficient condition for the sustainable management of the global commons without the internalization of environmental costs and benefits. Likewise, the insights gained in the second analysis can be translated to the international level. Property rights based approaches to the management of the global commons need to take into account the extent of collective action problems among appropriators from the commons and the relationship between the economic and environmental values of the resource. Furthermore, designers of such approaches need to critically evaluate capacity and commitment of the respective supra-national authorities charged with supervising the implementation and adjusting the design of the scheme if necessary.

6.3 The ideal stage: what really needs to be done

Leaving behind the world of the currently realistic and feasible, for a moment, what would be the ideal stage at which we ultimately want to arrive? We can address this question in terms of environmental governance as well as in terms of research. In the following discussion, each will be taken up in turn.

Starting with the academic side, it is obvious that the analyses in this book have predominantly relied on a neo-institutionalist approach. This approach clearly has strengths, in particular its ability to focus on the implications of incentives for individual decision-making and outcomes of interactive decision-making between groups of individuals. At the same time, the approach has obvious weaknesses as well. The assumption of given preferences as a basis for the deduction of incentives and decisions is one of the most controversial of these. In the analyses in this book, for instance, the general assumption has been that the individual has the preference structure of homo oeconomicus. In addition, the positivist view underlying neo-institutionalism can easily be questioned, especially in the context of environmental problems.

In political science, constructivist approaches have brought the most serious challenges against neo-institutionalism. If applied to the analyses in this book, constructivist criticism would point out that other preference structures besides homo oeconomicus exist, and that preferences change frequently. Every day numerous individuals decide to forego private economic benefit for collective environmental benefit, for instance. Likewise, constructivists would highlight that numerous other factors besides clear preference structures of individuals, such as norms and habits, influence individual decision-making and guide behavior. Moreover, constructivist would point out the lack of 'objective environmental problems.' All of these criticism relate back to one fundamental assumption in constructivist thinking: the social construction of our reality, i.e. the extent to which experience and cognitive mindsets influence our perception of nature, meaning, or social interaction (Kratochwil, 1989; Ruggie, 1998). In terms of environmental problems, Litfin (1999), for instance, has highlighted the ways in which we construct rather than measure environmental problems. The same dynamic applies to property rights. Social experience and cognitive frames of mind influence our perception of the meaning and nature of property rights as well as our choice sets in resource management.

In an ideal world, then, we would be able to learn something from both perspectives, neo-institutionalism and constructivism, if not integrate them.[1]

[1] Wendt (1999) has attempted to move towards an integration of tools and insights from different approaches for the area of international politics.

After all, both perspectives have strengths and weaknesses. However, the perspectives utilize such different assumptions and scientific approaches that integration is extremely difficult. In addition, scholars writing in each of the two perspectives often tend to fundamentally reject the other and deny any potential of integration. Both approaches allow valuable insights, however. With respect to the analyses at hand, for instance, cognitive mindsets and social norms clearly play a role in resource management, but so do economic incentives. Thus, in the ideal world, further research needs to figure out how to include the benefits of one perspective without excluding the other. Future progress not just in environmental governance and resource management but in political science as such will depend on our ability to learn from both.

To find an answer to 'the ideal stage' from a more applied agenda, i.e. in terms of environmental governance, it is necessary to draw the circle a bit wider. The links to the above answer on future research, however, should become clear very quickly. There are a significant number of scholars who would challenge the arguments made in this book on practical, philosophical and ethical grounds (Pearce and Warford, 1993; Taylor, B., 1992; Daly and Cobb, 1989; Sagoff, 1988). These scholars would suggest that with the reduction of natural resources to 'property' one advocates a commodification of nature which is not only a major cause of our current environmental problems, but is also a fiction. It is a fiction, in their view, because the application of property rights is based on the assumption that we can evaluate and own parts of nature separately from each other. These scholars emphasize that economic simplifications necessary for this commodification of nature force us to ignore the interdependence between natural resources, the intergenerational distribution of costs and benefits from natural resources as well as questions of irreversibility or non-monetarized values of natural resources, which are the core characteristics of nature.

One can challenge some of these criticisms by pointing out that it is not the institution of property rights that is failing us, but rather that we fail to attribute appropriate values to property rights over natural resources.[2] Property rights as an institution can take anything into account that we want them to take into account. It is our current tendency to be guided by economic incentives that induces us to ignore interdependencies and intergenerational aspects. If we attached a higher value to the welfare of future generations, for instance, we would not apply as a high discount rate to future benefits from resources as we obviously do.

It is true, then, that in the context of the current economic system with the zero value we attach to many natural resources, such as air and water, the assurance of property rights, for instance, will not take us sustainably into

[2] Deep ecologists will point out, of course that 'property,' by definition, excludes the notion of intrinsic value in what is owned.

the 22nd century. It is also true, however, that a sustainable economic system is not necessarily devoid of property rights. Thus, to critics arguing that the security of property rights has led to major environmental degradation in the United States from colonial times to the capitalist present (as well as in other developed countries), one could answer that the degradation occurred predominantly in resources over which no property rights were defined, or in resources for which the property rights were not appropriately valued.[3] I am not convinced that in the latter case natural resources would necessarily have fared better if no property rights had been defined, although I acknowledge that degradation may have been slower to arrive.[4]

Some critics would go further in their challenges of the analyses presented in this book, although they would arrive at a similar conclusion: the commodification of nature is a fiction and, as an instrument in the quest for wealth, the cause of environmental disaster (Cronon, 1995; Ekins et al., 1992). These critics would point out that the detrimental impact of property rights on environmental quality they perceive, or the inadequate valuation of rights associated with an environmentally sustainable use of natural resources, derive from a wrong worldview establishing a hierarchy between human- and non-human nature. The worldview, originating at the time and from the philosophy of Bacon and Descartes, asserts the capability and rights of the human race to be the master over nature. This mastery, in turn, supposedly bears fruits in terms of social benefits: "an increased supply of goods and a general liberation of the intellect from superstition and irrationality - that would enable men to control their desires and to pursue their mutual concerns more justly and humanely" (Leiss 1972: 21).[5] According to these presumed critics, this belief, then, has led us to value natural resources solely in terms of their perceived use for humankind, to a neglect of the inherent value of nature, and ultimately to its destruction.

Against this argument, it is difficult to defend oneself. Most scholars working on environmental issues see the benefits of an ecocentric

[3] Alternatively, it might be appropriate to argue that 'property rights' of Native Americans existed, but they were not recognized and respected by the colonists, although Native Americans would argue that the concept of 'property' was imposed on them.

[4] The negative impact of property rights on open-access resources would result from the positive impact of property rights on economic growth with the corresponding increased use of resources. While the assurance of property rights can also benefit the potential for environmental stewardship of open-access, this positive effect might not balance the negative effect of economic growth.

[5] This worldview has continued into the 20th century and can, for instance, be seen as the source of Pinchot's "unwavering optimism about the possibility of correcting this problem [the reckless waste and destruction of natural resources] and, in fact, managing natural resources so as to allow for an almost endlessly increasing American prosperity" (Taylor, B., 1992: 18.)

worldview. Yet, our reality is based on an anthropocentric worldview, and the question of whether we will ever be willing and able to move to an ecocentric one is still up for debate. In the meantime, therefore, we need to optimize the potential for environmental sustainability within the constraints of the current dominant Western economic paradigm. As such, the assurance of property rights is a valuable strategy, justified in terms of policy relevance as it allows improvements in the short-term. It is not sufficient, and, in the absence of an adequate valuation of environmental benefits, not an alternative to environmental governance.[6] Yet, the analyses in this book have shown that governments can set institutional frameworks inducing improvements in the potential for environmental stewardship.

[6] The reasons for this regulation may be ethical, cultural, or environmental, as much as economic (Sagoff, 1988). Moreover, I am not advocating the free market. In terms of externalities and other problems such as imbalances, the necessary internalization, for instance, can only be achieved by a collective process (Ekins et al., 1992). Finally, I am well aware that the assurance of property rights cannot overcome the damage done by other misguided government policies, such as those of the Brazilian government encouraging the destruction of the rain forest.

Bibliography

Adger, W. Neil, and Cecilia Luttrell (2000) "Property Rights and the Utilization of Wetlands." *Ecological Economics* 35: 75-89.

Alchian, A. (1967) "Pricing and Society." Westminster: The Institute of Economic Affairs. Occasional Paper No. 17.

Alesina, Alberto, Sule Ozler, Nouriel Roubini, and Philip Swagel (1991) "Political Instability and Economic Growth." Cambridge: NBER Research.

Alker, Hayward R. Jr. and Peter M. Haas (1993) "The Rise of Global Ecopolitics." in: Choucri, Nazli. *Global Accord. Environmental Challenges and International Responses*. Massachusetts Institute of Technology: Cambridge.

Allen, Douglas W. (1991) "What Are Transaction Costs?" *Research in Law and Economics* 14: 1-18.

Amacher, Ryan C., Robert D. Tollison, and Thomas D. Willett. (1972). "The Economics of Fatal Mistakes: Fiscal Mechanisms for Preserving Endangered Predators." *Public Policy* 20: 411-441.

Arbetman, Marina, and Jacek Kugler (eds.) (1997) *Political Capacity & Economic Behavior*. Boulder :Westview Press.

Arrow, Kenneth, B. Bolin, R. Constanza, P. Dasgupta, C. Folke, S. Holling, B.-O. Jansson, S. Levin, K.-G. Mäler, C. Perrings, D. Pimentel (1995) "Economic Growth, Carrying Capacity and the Environment." *Science* 268: 520-521.

Axelrod, Robert (ed.) (1976) *Structure Of Decision: The Cognitive Maps Of Political Elites*. Princeton: Princeton University Press.

Baden, John A., and Tim O'Brien (1997) "Bringing Private Management to the Public Lands: Environmental and Economic Advantages." in: Sheldon Kamieniecki, George Gonzalez, and Robert Vos (eds.) *Flashpoints in Environmental Policymaking*. Albany: State University of New York Press.

Bardaracco, Joseph (1985) *Loading the Dice. A Five-Country Study of Vinyl Chloride Regulation*. Boston: Harvard Business School Press.

Barro, Robert (1991. "Economic Growth in a Cross Section of Countries." *Quarterly Journal of Economics*. 61: 407-44.

Barkin, J. Samuel, and George E. Shambaugh (eds.) (1999) *Anarchy and the Environment: The International Relations of Common Pool Resources*. Albany: Suny Press.

Barzel, Yoram (1989) *Economic Analysis of Property Rights*. Cambridge: Cambridge University Press.

Beaumont, Paul M., and Robert T. Walker (1996) "Land Degradation and Property Regimes." *Ecological Economics* 18: 55-66.

Berge, Erling, and Nils Christian Stenseth (1998) *Law and the Governance of Renewable Resources*. Oakland: ICS Press.

Berkes, Fikret (1992) "Success and Failure in Marine Coastal Fisheries of Turkey." in: Daniel W. Bromley (ed.) *Making the Commons Work*. San Francisco: ICS Press, pp. 161-182.

Böhmer-Christiansen, Sonja, and Jim Skea (1991) *Acid Politics: Environmental and Energy Policies in Britain and Germany*. New York: Belhaven.

Böhmer-Christiansen, Sonja (1992) "Anglo-German Contrasts in Environmental Policy-Making and their Impacts in the Case of Acid Rain Abatement." *International Environmental Affairs* 4(4): 295-322.

Borner, Silvio, and Aymo Brunetti, and Beatrice Weder (1995) "Policy Reform and Institutional Uncertainty: The Case of Nicaragua." *KYKLOS* 48: 43-64.

Botkin, Daniel B., and Edward A. Keller (1995) *Environmental Science. Earth as a Living Planet*. New York: John Wiley & Sons, Inc..

Bressers, Hans, and Lawrence O'Toole (1998) "The selection of policy instruments: A network-based perspective." *Journal of Public Policy* 3(3): 213-239.

Brickman, Ronald, Sheila Jasanoff, and Thomas Ilgen (1985) *Controlling Chemicals: The Politics of Regulation in Europe and the United States*. Ithaca: Cornell University Press.

Bromley, Daniel W. (ed.) (1992) *Making the Commons Work. Theory, Practice, and Policy*. San Francisco: ICS Press.

Bromley, Daniel W. (1991) *Environment and Economy. Property Rights and Public Policy*. Cambridge: Blackwell.

Bromley, Daniel W. (1989) *Economic Interests and Institutions. The Conceptual Foundations of Public Policy*. New York: Basil Blackwell Inc.

Bromley, Daniel W. (1978) "Property Rules, and Environmental Economics." *Journal of Economic Issues* 12: 43-60.

Buchanan, James M. (1975) "Before Public Choice." in: Henry G. Manne (ed.) *The Economics of Legal Relationships. Readings in the Theory of Property Rights*. St. Paul, New York, a.o.: West Publishing Company. pp. 67-77.

Buell, John and Tom DeLuca (1996) *Sustainable Democracy: Individuality And The Politics Of The Environment*. Thousand Oaks: Sage Publications.

Burger, Joanna, and Michael Gochfeld (1998) "The Tragedy of the Commons at 30 Years." *Environment* 40(10): 4-13, 26-27.

Carraro, Carlo, and Domenico Siniscalco (1991) "Strategies for the International Protection of the Environment." *CEPR Discussion Paper No. 568*. London: Centre for Economic Policy Research.

Chakraborty, Rabindra Nath (2001) "Stability and Outcomes of Common Property Institutions in Forestry: Evidence from the Terai Region of Nepal." *Ecological Economics* 36: 341-353.

Cheung, Steven N.S. (1975) "The Structure of A Contract and the Theory of a Non-Exclusive Resource." in: Henry G. Manne (ed.) *The Economics of Legal Relationships. Readings in the Theory of Property Rights*. St. Paul, New York, a.o.: West Publishing Company. pp. 437-56.

Choucri, Nazli (1993) *Global Accord. Environmental Challenges and International Responses*. Massachusetts Institute of Technology: Cambridge.

Choucri, Nazli and Robert C. North (1993a) "Growth, Development, and Environmental Sustainability: Profiles and Paradox." in: Choucri, Nazli. *Global Accord. Environmental Challenges and International Responses*. Massachusetts Institute of Technology: Cambridge.

Choucri, Nazli, and Robert C. North (1993b) "Global Accord: Imperatives for the Twenty-First Century." in: Choucri, Nazli. *Global Accord. Environmental Challenges and International Responses*. Massachusetts Institute of Technology: Cambridge.

Clague, Christopher, Philip Keefer, Stephen Knack, and Mancur Olson (1995) "Contract-Intensive Money: Contract Enforcement, Property Rights, and Economic Performance." Unpublished Manuscript.

Coase, R.H. (1960) "The Problem of Social Cost." *The Journal of Law and Economics* 3: 1-44.

Cornell, Stephen, and Joseph P. Kalt (1995) "Where Does Economic Development Really Come From? Constitutional Rule among the Sioux and the Apache." *Economic Inquiry* 33: 402-426.

Cornes, Richard, and Todd Sandler (1983) " On Commons and Tragedies." *American Economic Review* 73: 787-92.

Costanza, Robert, and Herman E. Daly (1987) "Toward and Ecological Economics." *Ecological Modelling* 38: 1-7.

Cronon, William (ed.) (1995) *Uncommon Ground. Toward Reinventing Nature*. New York: W.W. Norton & Company.

Dalton, Russel J. (1994) *The Green Rainbow. Environmental Groups in Western Europe*. New Haven: Yale University Press.

Daly, Herman E., and John B. Cobb, Jr. (1989) *For the Common Good: Redirecting the Economy Toward Community, the Environment, and a Sustainable Future*. Boston: Beacon Press.

Dasgupta, Partha, and Karl-Göran Mäler. (1994). *Poverty, Institutions, and the Environmental-Resource Base*. World Bank Environment Paper Number 9. The World Bank: Washington D.C.

Dasgupta, Susmita, Ashoka Mody, Subhendy Roy, and David Wheeler (1995) *Environmental Regulation and Development: A Cross-Country Analysis*. Washington, D.C.: The World Bank.

Dauvergne, Peter (1997) *Shadows in the Forest: Japan and the Politics of Timber in Southeast Asia*. Boston: MIT Press.

Davis, Charles (1997) "This Land is "Our" Land: The Case for Federal Retention of Public Lands." in: Sheldon Kamieniecki, George Gonzalez, and Robert Vos (eds.) *Flashpoints in Environmental Policymaking*. Albany: State University of New York Press.

Deacon, Robert T. (1995) "Assessing the Relationship between Government Policy and Deforestation." *Journal of Environmental Economics and Management* 28: 1-18.

Deacon, Robert T. (1994) "Deforestation, Investment and Ownership Security." Unpublished Work.

De Alessi, L. (1991) "Development of the Property Rights Approach." in: Furubotn, Eirik G., and Rudolf Richter (eds.) *The New Institutional Economics. A Collections of Articles from the Journal of Institutional and Theoretical Economics*. College Station: Texas A&M University Press. pp. 45-53.

De Bruyn, Sander M. (1997) "Explaining the environmental Kuznets curve: structural change and international agreements in reducing sulphur emissions." *Environment and Developmental Economics* 2: 485-503.

De Bruyn, Sander M., J.C.J.M. van den Bergh, and J.B. Opschoor (1998) "Economic Growth and Emissions: Reconsidering the Empirical Basis of Environmental Kuznets Curves." *Ecological Economics* 25: 161-175.

De Bruyn, Sander M., and J.B. Opschoor (1997) "Developments in the Throughput-Income Relationship: Theoretical and Empirical Observations." *Ecological Economics* 20: 255-268.

Demsetz, Harold (1975a) "Toward a Theory of Property Rights." in: Henry G. Manne (ed.) *The Economics of Legal Relationships. Readings in the Theory of Property Rights*. St. Paul, New York, a.o.: West Publishing Company. pp. 23-36.

Demsetz, Harold (1975b) "When Does the Rule of Liability Matter?" in: Henry G. Manne (ed.) *The Economics of Legal Relationships. Readings in the Theory of Property Rights*. St. Paul, New York, a.o.: West Publishing Company. pp. 168-83.

Demsetz, Harold (1975c) "Some Aspects of Property Rights." in: Henry G. Manne (ed.) *The Economics of Legal Relationships. Readings in the Theory of Property Rights*. St. Paul, New York, a.o.: West Publishing Company. pp. 184-93.

Demsetz, Harold (1975d) "The Exchange and Enforcement of Property Rights." in: Henry G. Manne (ed.) *The Economics of Legal Relationships. Readings in the Theory of Property Rights*. St. Paul, New York, a.o.: West Publishing Company. pp. 362-77.

Denzau, Arthur, and Douglass North (1994) "Shared Mental Models: Ideologies and Institutions." *KYKLOS* 47: 3-31.

De Soto, Hernando (1989) *The Other Path*. New York: Harper & Row.

Devlin, Rose Anne, and Quentin Grafton (1998) *Economic Rights and Environmental Wrongs. Property Rights for the Common Good*. Cheltenham: Elgar.

Dinda, Soumyananda, Dipankor Coondoo, and Manoranjan Pal (2000) "Air Quality and Economic Growth: An Empirical Study." *Ecological Economics* 34: 409-423.

Dryzek, John (1997) *The Politics Of The Earth: Environmental Discourses*. New York: Oxford University Press.

Dryzek, John (1987) *Rational Ecology: Environment and Political Economy*. Oxford and New York: Basil Blackwell.

Eggertson, Thráinn (1998) "The Economic Rational of Communal Resources." in: Berge, Erling, and Nils Christian Stenseth (eds.) *Law and the Governance of Renewable Resources*. Oakland: ICS Press.

Eggertson, Thráinn (1996) "The Economics of Control and the Cost of Property Rights." in: Hanna, Susan, Carl Folke, and Karl-Göran Mäler (eds.) *Rights to Nature*. Washington, D.C.: Island Press.

Eliasson, Gunnar (1995) *Investment Incentive in the Formerly Planned Economies*. The Royal Institute of Technology, Department of Industrial Economics and Management: Stockholm.

Ekins, Paul (1997) "The Kuznets Curve for the Environment and Economic Growth: Examining the Evidence." *Environment_and_Planning* 29: 805-830.

Ekins, Paul, Mayer Hillman, and Robert Hutchinson (1992) *The Gaia Atlas of Green Economics*. New York: Anchor Books Doubleday.

Elkins, David J., and Richard E.B. Simeon (1979) "A Cause in Search of Its Effect, of What Does Political Culture Explain?" *Comparative Politics* 11: 127-45.

Emerson, Kirk, and Charles R. Wise (1995) "Statutory Approaches to Regulatory Takings. State Property Rights Legislation. Issues and Implications for Public Administration." Paper Prepared for the Annual Meeting of the American Political Science Association, Chicago, 1995.

Enloe, Cynthia (1975) *The Politics of Pollution in a Comparative Perspective. Ecology and Power in Four Nations*. New York: David McKay.

Ensminger, Jean, and Andrew Rutten (1991) "The Political Economy of Changing Property Rights: Dismantling a Pastoral Commons." *American Ethnologist* 18(4): 41-57.

European Commission, DG XI (1998) "Workshop on Sustainable Development - Challenge for the Financial Sector." *Final Report*. Brussels: The European Commission.

European Commission, DG XI (1997) "The Role of Financial Instruments in Achieving Sustainable Development." *Report*. Brussels: The European Commission.

Feeny, David (1998) "Suboptimality and Transaction Costs on the Commons." in: Loehman, Edna T., and D. Marc Kilgour (eds.) *Designing Institutions for Environmental and Resource Management*. Cheltenham: Edward Elgar.

Feeny, David (1988) "The Development of Property Rights in Land: A Comparative Study." in: Robert H. Bates. *Toward a Political Economy of Development. A Rational Choice Perspective*. Berkeley, Los Angeles, and London: University of California Press.

Feeny, David, Fikret Berkes, Bonnie J. McCay, and James M. Acheson (1990) "The Tragedy of the Commons: Twenty-Two Years Later." *Human Ecology* 18(1): 1-18.

Fisher, Frank (1995) *Evaluating Public Policy*. Chicago: Nelson-Hall.

Fisher, Frank, and John Forrester (eds.) (1993) *The Argumentative Turn In Policy Analysis And Planning*. Durham: Duke University Press.

Furubotn, Eirik G., and Rudolf Richter (eds.) (1991) *The New Institutional Economics. A Collections of Articles from the Journal of Institutional and Theoretical Economics.* College Station: Texas A&M University Press.

Furubotn, Eirik G., and Svetozar Pejovich (1975) "Property Rights and Economic Theory: A Survey of Recent Literature." in: Henry G. Manne (ed.) *The Economics of Legal Relationships. Readings in the Theory of Property Rights*. St. Paul, New York, a.o.: West Publishing Company. pp. 53-65.

Gallup Poll (1989) "The Environment." *The Gallup Report* 285.

Gangadharan, Lata, and Ma. Rebecca Valenzuela (2000) "Interrelationships between Income, Health, and the Environment: Extending the Environmental Kuznets Curve Hypothesis." *Ecological Economics* 36(3): 513-31.

Goldsmith, Edward, et al. (1972) *Blueprint for Survival*. Boston: Houghton Mifflin.

Gordon, Scott H. (1954) "The Economic Theory of a Common-Property Resource: the Fishery." *Journal of Political Economy* 62: 124-42.

Gore, Albert, Jr. (1992) *Earth in the Balance: Ecology and the Human Spirit*. New York: Houghton Mifflin.

Greene, William H. (1993) *Econometric Analysis*. New York: Macmillan Publishing Company.

Grossman, Gene M., and Alan B. Krueger (1995) "Economic Growth and the Environment." *Quarterly Journal of Economics* pp. 353-377.

Grossman, Gene M., and Alan B. Krueger (1994) "Economic Growth and the Environment." NBER Working Paper No. 4634. Cambridge: National Bureau of Economic Research.

Grossman, Gene M., and Alan B. Krueger (1993) "Environmental Impacts of a North American Free Trade Agreement," in: P. Garber, (ed.) *The U.S.-Mexico Free Trade Agreement*. Cambride: MIT Press.

Grossman, Gene M., and Alan B. Krueger. (1991) "Environmental Impacts of a North American Free Trade Agreement." Paper prepared for the SECOFE conference on the U.S.-Mexico Free Trade Agreement.

Gurr, Ted Robert, Keith Jaggers, and Will H. Moore (1991) "The Transformation of the Western State: The Growth of Democracy, Autocracy, and State Power since 1800." in: Alex Inkeles (ed.) *On Measuring Democracy. Its Consequences and Concmitants*. New Brunswick: Transaction Publishers.

Hanna, Susan, Carl Folke, and Karl-Göran Mäler (1996) *Rights to Nature*. Washington, D.C.: Island Press.

Hardin, Garrett (1968) "The Tragedy of the Commons." *Science* 162: 1243-1248.

Hayward, Bronwyn M. (1995) "Beyond Liberalism? Environmental Management and Deliberative Democracy." Paper presented at the 91st Annual Meeting of the American Political Science Association, Chicago, Illinois.

Heemskerk, K.N. (1997) "Sustainable Development: A Challenge for Banks?" Dissertation, ERASMUS University, Faculty of Business Administration.

Heilbroner, Robert L. (1980) *An Inquiry into the Human Prospect. Updated and Reconsidered for the 1980*. New York: Norton and Company.

Heilbroner, Robert L. (1975) *An Inquiry into the Human Prospect*. New York: W.W. Norton.

Hempel, Lamont C. (1996) *Environmental Governance: The Global Challenge*. Washington, D.C.: Island Press.

Hirschman, Albert O. (1970) *Exit, Voice, and Loyality - Responses to Decline in Firms, Organizations, and States*. Cambridge: Harvard University Press.
Hobbes, Thomas (1991) *Leviathan*. New York: Cambridge University Press.
Homer-Dixon, Thomas F. (1993) "Physical Dimensions of Global Change." in: Choucri, Nazli. *Global Accord. Environmental Challenges and International Responses*. Massachusetts Institute of Technology: Cambridge.
Hsiang, Cheng (1986) *Analysis of Panel Data*. New York: Cambridge University Press.
Inglehart, Ronald (1990) *Culture Shift in Advanced Industrial Societies*. Princeton: Princeton University Press.
Inglehart, Ronald (1971) "The Silent Revolution in Europe: Intergenerational Change in Postindustrial Societies." *American Political Science Review* 65: 991-1007.
Inglehart, Ronald, and Paul R. Abramson (1994) "Economic Security and Value Change." *American Political Science Review* 88(2): 336-353.
Jänicke, Martin, Manfred Binder, Stefan Bratzel, Alexander Carius, Helge Joergens, Kristine Kern, and Harald Mönch (1995) *Umweltpolitik im internationalen Vergleich: Untersuchungen zu strukturellen Erfolgsbedingungen*. Forschungsstelle fuer Umweltpolitik: Berlin.
Jänicke, Martin, and Harald Mönch (1988) "Ökologischer und Wirtschaftlicher Wandel im Industrieländervergleich. Eine explorative Studie über Modernisierungskapazitäten." in: Manfred G. Schmidt (ed.) *International und Historisch Vergleichende Analysen*. PVS Sonderheft 19. Opladen: Westdeutscher Verlag.
Kaufmann, Robert, Brynhildur Davidsdottir, Sophie Garnham, Peter Pauly (1998) "The determinants of atmospheric SO2 concentrations: reconsidering the environmental Kuznets curve." *Ecological Economics* (25) 2: 209-220.
Keefer, Philip, and Stephen Knack (1994) "Property Rights, Inequality and Growth." Unpublished Manuscript.
Keohane, Robert and Elinor Ostrom (1995) *Local Commons and Global Interdependence*. Thousand Oaks: SAGE.
Kissling-Näf, Ingrid, and Frédéric Varone (eds.) (2000) *Institutionen für eine nachhaltige Ressourcennutzung. Innovative Steuerungsansätze am Beispiel der Ressourcen Luft und Boden*. Zürich: Rüegger.
Kitschelt, Herbert P. (1986) "Political Opportunity Structures and Political Protest: Anti-Nuclear Movements in Four Democracies." *British Journal of Political Science* 16(1): 57-85.
Kobler, Markus (2000) *Der Staat und die Eigentumsrechte*. Tübingen: Mohr Siebeck.
Knöpfel, Peter, and Helmut Weidner (1985) *Luftreinhaltepolitik (Stationäre Quellen) im Internationalen Vergleich*. Berlin: edition sigma.
Kraft, Michael. E. (1996) *Environmental Policy and Politics: Toward the Twenty-First Century*. New York: Harper Collins.
Kratochwil, Friedrich (1989) *Rules, Norms, and Decisions*. Cambridge: Cambridge University Press.
Leal, Donald (1996) "Community-Run Fisheries: Avoiding the "Tragedy of the Commons"." *PERC Policy Series* PS-7.
Leiss, William (1972) *The Domination of Nature*. New York: George Braziller.
Levi, Margaret (1988) *Of Rule and Revenue*. Berkeley: University of California Press.
Libecap, Gary (1998) "Distributional and Political Issues in Modifying Traditional Common-property Institutions." in: Berge, Erling, and Nils Christian Stenseth (eds.) *Law and the Governance of Renewable Resources*. Oakland: ICS Press.
Libecap, Gary (1991) "Distributional Issues in Contracting for Property Rights." in: Eirik G. Furubotn and Rudolf Richter (eds.) *The New Institutional Economics. A Collection of*

Articles from the Journal of Institutional and Theoretical Economics. College Station: Texas A&M University Press. pp. 214-32.

Litfin, Karen (1999) "Constructing Environmental Security and Ecological Interdependence." *Global Governance* 5: 359-377.

Livingston, Marie Leigh (1987) "Evaluating the Performance of Environmental Policy: Contributions of Neoclassical, Public Choice, and Institutional Models." *Journal of Economic Issues* 21: 281-295.

Loehman, Edna T., and D. Marc Kilgour (1998) *Designing Institutions for Environmental and Resource Management*. Cheltenham: Edward Elgar.

Lovins, Amory (1977) *Soft Energy Paths*. Cambridge: Ballinger.

Lowry, William (1998) *Preserving Public Lands for the Future. The Politics of Intergenerational Goods*. Washington, D.C.: Georgetown University Press.

Lueck, Dean (1995) "Property Rights and the Economic Logic of Wildlife Institutions." *Natural Resources Journal* 35: 625-670.

Lundquist, Lennart (1974) "Do Political Structures Matter in Environmental Politics? The Case of Air Pollution Control in Canada, Sweden, and the United States." *Canadian Public Administration* 17: 119-141.

Lundquist, Lennart (1980) *The Hare and the Tortoise: Clean Air Policies in the United States and Sweden*. Ann Arbor: University of Michigan Press.

Mansbridge, Jane (1980) *Beyond Adversary Democracy*. New York: Basic Books.

Margerum, Christine (1995). "The Decentralization of Policymaking: Lessons from the U.S. and German Subnational Support for High-Technology Industries." Paper presented at the 91st Annual Meeting of the American Political Science Association, Chicago, Illinois.

Martin, Fenton (1992) *Common Pool Resources and Collective Action: A Bibliography*. Vol. 2. Bloomington: Indiana University, Workshop in Political Theory and Policy Analysis.

Mazmanian, Daniel A. (1995) "Clear Vision, Clean Skies: A New Epoch in Air Pollution Control for the Los Angeles Region." Monograph. Claremont: The Center for Politics and Economics, The Claremont Graduate School.

Mazmanian, Daniel A., and David Morell (1990) "The NIMBY Syndrome: Facility Siting and the Failure of Democratic Discourse." in: Norman J. Vig, and Michael E. Kraft. (eds) *Environmental Policy in the 1990s*. Washington, D.C.: Congressional Quarterly Press.

McCay, Bonnie J., and James M. Acheson (eds.) (1987) *The Question of the Commons. The Culture and Ecology of Communal Resources*. Tucson: The University of Arizona Press.

McChesney, Fred S. (1990) "Government as Definer of Property Rights: Indian Lands, Ethnic Externalities, and Bureaucratic Budgets." *Journal of Legal Studies* 19: 297-335.

McGrory Klyze, Christopher (1996) *Who Controls Public Lands?* Chapel Hill: The University of North Carolina Press.

McIntyre, Robert J. and James R. Thornton (1978) "On the Environmental Efficiency of Economic Systems." *Soviet Studies* Vol. XXX(2): 173-92.

McKean, Margaret (1992) "Management of Traditional Common Lands (Iriaichi) in Japan." in: Daniel W. Bromley (ed.) *Making the Commons Work*. San Francisco: ICS Press.

Midlarsky, Manus (1998) "Democracy and the Environment: An Empirical Assessment." *The Journal of Peace Research* 35(3): 341-361.

Milbrath, Lester (1993) "The World Is Relearning Its Story About How The World Works." in: Sheldon Kamieniecki (ed.) *Environmental Politics In The International Arena*. Albany: SUNY.

Mohr, Ernst (1990) "Courts of Appeal, Bureacracies and Conditional Project Permits: The Role of Negotiating Non-Exclusive Property Rights over the Environment." *Journal of Institutional and Theoretical Economics* 146: 601-616.

Moomaw, W.R., and G.C. Unruh, W.R. (1998) "An alternative analysis of apparent EKC-type transitions." *Ecological Economics* (25)2: 221-229.
Mueller-Rommel, Ferdinand (1993) *Grüne Parteien in Westeuropa. Entwicklungsphasen und Erfolgsbedingungen.* Opladen: Westdeutscher Verlag.
National Research Council (1986) *Proceedings of the Conference on Common Property Resource Management.* Washington, D.C.: National Academy Press.
Naughton-Treves, Lisa, and Steven Sanderson (1995) "Property, Politics, and Wildlife Conservation." *World Development* 23(8): 1265-75.
Nelkin Dorothy and Michael Pollack (1981) *The Atom Besieged. Extraparliamentary Dissent in France and Germany.* Cambridge: MIT Press.
North, Douglass (1991a) "Transaction Costs, Institutions, and Economic History." in: Eirik G. Furubotn and Rudolf Richter (eds.) *The New Institutional Economics. A Collection of Articles from the Journal of Institutional and Theoretical Economics.* College Station: Texas A&M University Press. pp. 203-13.
North, Douglass (1991b) "A Transaction Cost Approach to the Historical Development of Polities and Economies." in: Eirik G. Furubotn and Rudolf Richter (eds.) *The New Institutional Economics. A Collection of Articles from the Journal of Institutional and Theoretical Economics.* College Station: Texas A&M University Press. pp. 253-60.
North, Douglass (1989) "Institutions and Economic Growth: An Historical Introduction." *World Development* 17(9): 1319-1332.
North, Douglass, and Robert Thomas (1973) *The Rise of the Western World.* Cambridge: Cambridge University Press.
Olson, Mancur (1992) "Foreword." in: Todd Sandler. *Collective Action. Theory and Applications.* Ann Arbor: The University of Michigan Press.
Olson, Mancur (1965) *The Logic of Collective Action.* Cambridge: Harvard University Press.
Ophuls, William (1977) *Ecology and the Politics of Scarcity: Prologue to a Political Theory of the Steady State.* San Francisco: W.H. Freeman.
Ophuls, William and A. Stephen Boyan, Jr. (1992) *Ecology and the Politics of Scarcity Revisited. The Unraveling of the American Dream.* New York: W.H. Freeman and Company.
Orr, David W. (1992) *Ecological Literacy. Education and the Transition to a Postmodern World.* Albany: State University of New York Press.
Orr, David W. and Stuart Hill (1978) "Leviathan, the Open Society, and the Crisis of Ecology." *The Western Political Quarterly* 31(4): 457-69.
Ostrom, Elinor (1999) "Institutional rational choice: An assessment of the institutional analysis and development framework." in: Paul A. Sabatier (ed.) *Theories Of The Policy Process.* Boulder: Westview Press.
Ostrom, Elinor (1998) "A Behavioral Approach to the Rational Choice Theory of Collective Action." *American Political Science Review* 92(1): 1-22.
Ostrom, Elinor (1992) "The Rudiments of a Theory of the Origins, Survival, and Performance of Common-Property Institutions." in: Daniel W. Bromley (ed.) *Making the Commons Work. Theory, Practice, and Policy.* San Francisco: ICS Press.
Ostrom, Elinor (1990) *Governing the Commons. The Evolution of Institutions for Collective Action.* Cambridge: Cambridge University Press.
Ostrom, Elinor, Roy Gardner, and James Walker (1994) *Rules, Games, and Common Pool Resources.* Ann Arbor: University of Michigan Press.
Ostrom, Elinor, James Walker and Roy Gardner (1990) "Sanctioning by Participants in Collective Action Problems." Paper delivered at the Annual Meeting of the American Political Science Association, San Francisco, California.

Ozler, Sule, and Dani Rodrik (1992) "External Shocks, Politics, and Private Investment: Some Theory and Empirical Evidence." *NBER Working Paper* No. 3960. Cambridge.

Passmore, J. (1974) *Man's Responsibility for Nature*. New York: Basic Books.

Payne, Roger A. (1995) "Freedom and the Environment." *Journal of Democracy* 6(3): 41-55.

Pearce, David W., and Jeremy J. Warford (1993) *World Without End: Economics, Environment, and Sustainable Development*. New York: Oxford University Press.

Pejovich, Svetozar (1975) "Towards an Economic Theory of the Creation and Specification of Property Rights." in: Henry G. Manne (ed.) *The Economics of Legal Relationships. Readings in the Theory of Property Rights*. St. Paul, New York, a.o.: West Publishing Company. pp. 37-52.

Persson, Torsten, and Guido Tabellini (1994) "Is Inequality Harmful for Growth?" *American Economic Review* 84(3): 600-62.

Press, Daniel (1998) "Local Environmental Policy Capacity: A Framework for Research." *Natural Resources Journal* 38(1): 29-52.

Press, Daniel (1994) *Democratic Dilemmas in the Age of Ecology. Tress and Toxics in the American West*. Durham and London: Duke University Press.

Princen, Thomas (1998) "From Property Regime to International Regime: An Ecosystems Perspective." *Global Governance* 4(4): 395-413.

Reich, Michael R. (1984) "Mobilizing for Environmental Policy in Italy and Japan." *Comparative Politics* 16: 379-402.

Repetto, Robert (1988) "Overview." in: Robert Repetto and M. Gills. *Public Policies and the Misuse of Forest Resources*. Cambridge: Cambridge University Press.

Rinquist, Evan (1993) *Environmental protection at the State Level. Politics and Progress in Controlling Pollution*. Armonk: Sharpe.

Rinquist, Evan (1993) "Does Regulation Matter?: Evaluating the Effects of State Air Pollution Control Programs." *The Journal of Politics* 55(4): 1022-45.

Rosenthal, Jean-Laurent (1990) "The Development of Irrigation in Provence, 1700-1860: The French Revolution and Economic Growth." *The Journal of Economic History* L(3): 615-638.

Rothman, Dale, and Sander M. de Bruyn (1998) "Probing into the Environmental Kuznets Curve Hypothesis." *Ecological Economics* 25: 143-145.

Ruggie, John (1998) *Constructing the World Polity: Essays on International Institutionalization*. New York: Routledge.

Runge, C. Ford (1992) "Common Property and Collective Action in Economic Development." in: Daniel W. Bromley (ed.) *Making the Commons Work. Theory, Practice, and Policy*. San Francisco: ICS Press.

Sabatier, Paul (1988) "An Advocacy Coalition Framework Of Policy Change And The Role Of Policy-Oriented Learning Therein." *Policy Sciences* 21: 129-168.

Sabatier, Paul and Hank Jenkins-Smith (1999) "The Advocacy Coalition Framework: An Assessment." in: Paul Sabatier (ed.) *Theories Of The Policy Process*. Boulder: Westview Press.

Sabatier, Paul and Hank Jenkins-Smith (1993) *Policy Change And Learning: An Advocacy Coalition Approach*. Boulder: Westview Press.

Sagasti, Francisco R. and Michael E. Colby (1993) "Eco-Development and Perspectives on Global Change from Developing Countries." in: Choucri, Nazli. *Global Accord. Environmental Challenges and International Responses*. Massachusetts Institute of Technology: Cambridge.

Sagoff, Mark (1988) *The Economy of the Earth. Philosophy, Law, and the Environment*. Cambridge: Cambridge University Press.

Sandler, Todd (1992) *Collective Action. Theory and Applications*. Ann Arbor: The University of Michigan Press.

Sandler, Todd, and Frederic P. Sterbenz (1990) "Harvest Uncertainty and the Tragedy of the Commons." *Journal of Environmental Economics and Management* 18: 155-167.

Sanderson, Steven (1995) "Allocating Rights to Biota: Intellectual Property and the Preservation of Biodiversity." Paper presented at the Annual Meeting of the American Political Science Association, Chicago.

Schaltegger, Stefan, and Frank Figge (1998) "Environmental Shareholder Value." *WWZ/Sarasin Basic Report # 54*. Basel: WWZ, University of Basel.

Scharpf, Fritz (1997) *Games Real Actors Play: Actor-Centred Institutionalism In Policy Research*. Boulder: Westview Press.

Scheberle, Denise (1997) *Federalism and Environmental Policy: Trust and the Politics of Implementation*. Washington, D.C.: Georgetown University Press.

Schelbert-Syfrig, Heidi, and Andreas J. Zimmermann (1988) "Konkurrenz und Umweltschutz. Wald- und Holzwirtschaft zwischen Ökonomie und Ökologie." *Schweizerische Zeitschrift für Volkswirtschaft und Statistik* 3: 289-302.

Schlager, Edella, and William Blomquist (1998) "Heterogeneity and Common Pool Resource Management." in: Loehman, Edna T., and D. Marc Kilgour (eds.) *Designing Institutions for Environmental and Resource Management*. Cheltenham: Edward Elgar.

Schlager, Edella and Elinor Ostrom (1992) "Property-Rights Regimes and Natural Resources: A Conceptual Analysis." *Land Economics* 68(3): 249-62.

Schmidheiny, Stephan, and Federico Zorraquin (1996) *Financing Change. The Financial Community, Eco-Efficiency, and Sustainable Development*. Cambridge: The MIT Press.

Schön, Donald, and Martin Rein (1994) *Frame Reflection: Toward The Resolution Of Intractable Policy Controversies*. New York: Basic Books.

Schrama, Geerten (1998) "Banks as change agents towards more sustainable industry?" Paper presented at the 7th International Conference of the Greening of Industry Network, Rome, Italy, November 15-18.

Schrama, Geerten, and Ferd Schelleman (1996) "Banks as External Stakeholders to Corporate Environmental Management. Trends in the Netherlands." *CSTM Studies and Reports # 45*. Enschede: CSTM, University of Twente.

Schumacher, E.F. (1973) *Small Is Beautiful: Economics as if People Mattered*. New York: Harper and Row.

Sedjo, Roger A. (1989) "Property Rights for Plants." *RESOURCES* 97: 1-4.

Selden, R. M., and D. Song (1994) "Environmental Quality and Development: Is there a Kuznets Curve for Air Pollution?" *Journal of Environmental Economics and Environmental Management* 27: 147-162.

Serageldin, Ismail (1995) "Sustainability and the Wealth of Nations: First Steps in an Ongoing Journey." Paper for the Third Annual World Bank Conference on Environmentally Sutainable Development. Preliminary Draft for Discussion Only.

Shafik, Nemat, and Sushenjit Banyopadhyay (1992) "Economic Growth And Environmental Quality. Time-Series and Cross-Country Evidence." Background Paper for World Development Report 1992. Washington, D.C.: The World Bank.

Shiva, Vandana (1999) "Diversity and Democracy: Resisting the Global Economy." *Global-Dialogue* 1(1): 19-30.

Shleifer, Andrei (1995) "Establishing Property Rights." in: *Proceedings of the World Bank Annual Conference on Development Economics 1994*. The International Bank for Reconstruction and Development/The World Bank.

Skolnikoff, Eugene B. (1993) "Science and Technology: The Sources of Change." in: Choucri, Nazli. *Global Accord. Environmental Challenges and International Responses.* Massachusetts Institute of Technology: Cambridge.

Solesbury, William (1976) "Issues and Innovation in Environmental Policy in Britain, West Germany, and California." *Policy Analysis* 2: 1-38.

Southgate, Douglas, John Sanders, and Simeon Ehui (1990) "Resource Degradation in Africa and Latin America: Population Pressure, Policies, and Property Arrangements." *American Journal of Agricultural Economics* 72(5): 1259-63.

Southgate, Douglas, Rodrigo Sierra, and Lawrence Brown (1991) "The Causes of Tropical Deforestation in Ecuador: A Statistical Analysis." *World Development* 19(9): 1145-51.

Span, Henry A. (1996) "Protecting Property from Democracy: Political Inequality and "Liberal" and "Civic Republican" Approaches to the Takings Clause." Paper Presented at the 1996 Annual Meeting of the American Political Science Association, San Francisco, California.

Stern, David, Michael S. Common, and Edward Barbier (1996) "Economic Growth and Environmental Degradation: The Environmental Kuznets Curve and Sustainable Development." *World Development* 24(7): 1151-1160.

Stubblebine, Wm. Craig (1975) "On Property Rights and Institutions." in: Henry G. Manne (ed.) *The Economics of Legal Relationships. Readings in the Theory of Property Rights.* St. Paul, New York, a.o.: West Publishing Company. pp. 11-22.

Suri, Vivek, and Duane Chapman (1998) "Economic growth, trade and energy: implications for the environmental Kuznets curve." *Ecological Economics* (25)2: 195-208.

Taylor, Bob Pepperman (1992) *Our Limits Transgressed. Environmental Political Thought in America.* Lawrence: University of Kansas Press.

Taylor, Michael (1987) *The Possibility of Cooperation.* Cambridge: Cambridge University Press.

Thirgood, J.V. (1981) *Man and the Mediterranean Forest: A History of Resource Depletion.* London: Academic Press.

Thompson, Michael, Richard Ellis, and Aaron Wildavsky (1990) *Cultural Theory.* Boulder: Westview Press.

Thomson, James T., David Feeny, and Ronald J. Oakerson (1992) "Institutional Dynamics: The Evolution and Dissolution of Common-Property Resource Management." in: Daniel W. Bromley (ed.) *Making the Commons Work.* San Francisco: ICS Press. pp. 129-160.

Torras, Mariano, James K. Boyce (1998) "Income, inequality, and pollution: a reassessment of the environmental Kuznets Curve." *Ecological Economics* (25)2: 147-160.

Torstensson, Johan (1994) "Property Rights and Economic Growth: An Empirical Study." *KYKLOS* 47: 231-247.

Turvey, Ralph (1975) "On Divergences between Social and Private Cost." in: Henry G. Manne (ed.) *The Economics of Legal Relationships. Readings in the Theory of Property Rights.* St. Paul, New York, a.o.: West Publishing Company. pp. 357-361.

United Nations Population Fund (1991) *Population Policies and Programmes: Lessons Learned from Two Decades of Experience.* New York: New York University Press.

United Nations Environmental Programme. "Public and Leadership Attitudes to the Environment in Four Continents. A Report of a Survey in 16 Countries." New York: Louis Harris and Associates.

Vogel, David (1993) "Representing Diffuse Interests in Environmental Policymaking." in: R. Kent Weaver and Bert A. Rockman (eds.) *Do Institutions Matter? Government Capabilities in the United States and Abroad.* Washington, D.C.: The Brookings Institution.

Weber, Edward P. (1998) *Pluralism by the Rules: Conflict and Cooperation in Environmental Regulation*. Washington, D.C.: Georgetown University Press.

Wendt, Alexander (1999) *Social Theory of International Politics*. Cambridge: Cambridge University Press.

White, Mark (1995) "Environmental Finance: Raising and Using Funds in an Age of Ecology." Paper presented at the 4th International Conference of the Greening of Industry Network, Toronto, Canada, November.

Wiebe, Keith, Abebayehu Tegene, and Betsey Kuhn (1997) "Managing Public and Private Land through Partial Interests." *Contemporary Economic Policy* XV: 35-43.

Williams, Burce A., and Albert R. Matheny (1995) *Democracy, Dialogue, and Environmental Disputes*. New Haven: Yale University Press.

World Bank (1995) *Monitoring Environmental Progress. A Report on Work in Progress*. The World Bank: Washington, D.C.

World Commission on Environment and Development (Brundtland Commission) (1987). *Our Common Future*. New York: Oxford University Press.

World Resources Institute (1994) *World Resources 1994-1995. People and the Environment*. New York: Oxford University Press.

World Resources Institute (1992) *World Resources 1992-1993. Toward Sustainable Development*. New York: Oxford University Press.

Young, Oran B. (1994) *International Governance. Protecting the Environment in a Stateless Society*. Cornell University Press: Ithaca and London.

Ziegler, Charles E. (1980) "Soviet Environmental Policy and Soviet Central Planning: A Reply to McIntyre and Thornton." *Soviet Studies* XXXII(1): 124-34.

Appendix A

As discussed in chapter 2, numerous quantitative analyses have been conducted in this area of research. These analyses tend to face substantial difficulties, for instance in terms of data availability and quality, questions of methodology, and problems with the inclusion of ecological aspects such as ecosystem resilience or irreversibility of damage. As a consequence, quantitative analyses are not at the center of the analyses in this book. However, some quantitative inquiries into the influence of institutions on environmental quality at a given level of economic development were conducted. The results of these analyses are presented here for those interested in further pursuing such assessments. Admittedly, these econometric analyses focus only on a limited group of the various institutional, social, and cognitive determinants of environmental quality identified by scholars. Besides the aim to concentrate on the most interesting ones, the selection was limited by measurability of the factors and data availability.

A.1 The dependent variable: environmental quality

The analysis follows the examples of Grossman and Krueger (1995), Shafik and Bandyopadhyay (1992), and Jänicke et al. (1995), in its choice of indicators of environmental quality. Data were obtained from Gene Grossman and Alan Krueger at Princeton, who in turn, drew the information directly from the Global Environmental Monitoring Systems (GEMS) project, and from the World Resources Institute, the World Bank, and the OECD. The 11 indicators of environmental quality range from measures of air quality in the form of concentrations of sulfur dioxide (SO_2) and light and heavy suspended particulate matter (smoke and SPM), of water quality in the form of dissolved oxygen (DO_2), biochemical oxygen demand (BOD), and fecal coliform in rivers, deforestation, and rural and urban access to safe drinking water and sanitation.

Indicators of Air Quality:
Sulfur Dioxide and Suspended Particulate Matter

Sulfur dioxide is a major form of air pollution today. It is a colorless and odorless gas, which is normally present at the Earth's surface at low concentrations, but has been dramatically increased through major anthropogenic sources, such as the burning of fossil fuels (mostly coal in power plants), as well as through a variety of industrial processes (petroleum refining, production of paper, cement and aluminum). In the atmosphere, complex interaction with oxygen and water vapor transforms SO_2 into sulfuric acid, which then is deposited as acid precipitation in the form of rainfall, snow, and fog. The impact of sulfuric acids depends on the dose or concentration present, which in turn is influenced by geology and climatic patterns, types of vegetation and soil composition, and ranges from severe damage to human and animal lung function to death for animals and plants, as well as corrosion of paint, metals and other building materials. Acid rain is responsible for the death of forests and lake ecosystems, as well as the acidification of soils.

Previous research has found SO_2 to be significantly correlated with per capita incomes. Specifically, a general agreement exists that the relationship between SO_2 and per capita income levels takes the form of an Environmental Kuznets Curve. However, Jänicke et al. (1995) point out that in their study the relationship between income levels and the SO_2 ratio is not very robust, as it almost disappears if you only look at the OECD countries without Turkey and Spain (which, they acknowledge, might be attributed to distortions in the design of their indicator). The reason for this relationship probably is that relatively 'simple' answers to SO_2 emissions are available. From the original idea to build higher smokestacks (which of course only turned a local into a regional problem), to technology to clean up coal so that it burns more cleanly, to the possibility to switch from coal with high sulfur content to coal with low sulfur content, to scrubbing or processes of coal gasification or fluidized-bed combustion, technological solutions have been found. In addition, policy advances have introduced economic incentives to reduce SO_2 emissions in the United States, for instance, in the form of the emissions trading program under the Clean Air Act. Finally, Jänicke et al. (1995) point out that high income levels might provide a relief from environmental problems in this regard apart from policy achievements, as they allow adjustments in the structure of fossil fuel consumption.

The dependent variable is measured as the median daily concentration of SO_2 ($\mu g/m^3$). Furthermore, the data document the urban or suburban location of the station, the nearby land use, and the type of measuring device used, which have an impact on the SO_2 concentrations measured. In addition, three control variables devised by Grossman and Krueger, capturing the population density, proximity to the coast, and volcanic eruptions nearby are utilized.[1] The sample, then, includes measurements from 228 stations in 40 countries between 1977 and 1991. As not all of these stations reported measurements in every year, the sample is of an unbalanced nature.

Suspended particulate matter forms the second indicator of air quality. Following Grossman and Krueger's lead, the analysis differentiates between heavy particulates and light particulates (smoke). SPM generally refers to small particles of solid or liquid substances, some of which are visible as smoke, soot, or dust, while others are not. Particulates affect human and animal health, and local ecosystems as well as the global one. In humans, particulates can cause chronic health problems, as they enter (and stay in) the lungs. Asbestos particles as well as particles of heavy metals (arsenic, copper, lead, zinc) that are emitted from industrial facilities such as smelters are particularly dangerous in this regard. Dust layers on plants affect their ability to absorb light, carbon dioxide, and water as well as the release of oxygen. The sources of SPM are many. They include modern agriculture, desertification, nearly all industrial processes, tobacco smoke, automobile exhaust, cement and coal dust, as well as a range of natural sources such as pollen, volcanic eruptions, fires, dust storms, human hair, sea spray, and beach sand. Their different characteristics (point sources, fugitive sources, mobile sources) imply correspondingly different potential for and costs of air quality improvement.

Previous research has analyzed the relationship between SPM and per capita income levels. Similar to their results for SO_2, Grossman and Krueger (1991, 1993, 1994, 1995) found evidence of an Environmental Kuznets Curve for smoke, but a uniform decrease in heavy particulates with rising per capita incomes. Likewise, Shafik and Bandyopadhyay (1992) report that SPM concentrations appear to increase with initial rises in per capita income levels, but fall with further increases once ascertain income level is reached. In contrast, Jänicke et al. (1995) find a negative linear relationship, indicating a decrease in SPM emissions with rising per capita income levels,

[1] As Grossman and Krueger point out, the inclusion of these control variables is not necessary to obtain unbiased results, since they are unlikely to be correlated with per capita income levels (and in this case, also with institutional capacity). They reduce the variance of the residuals, however, and thereby allow more precise estimates of the relationship between the indicators of environmental quality and institutional capacity and per capita income levels.

which might be due to the comparatively high income levels of the countries in their sample.

Again, the median daily concentrations of heavy particulate matter[2] and smoke are used as dependent variables ($\mu g/m^3$), and variables capturing land use, closeness to the coast, desert, population density and measuring devices as control variables. The sample includes measurements for heavy particulates from 152 stations in 25 countries and for smoke from 84 stations in 17 countries, from 1977-1991, and is again of an unbalanced nature.

Indicators of Water Quality:
Dissolved Oxygen (DO_2), Biochemical Oxygen Demand (BOD), and Fecal Coliform

One indicator of water quality in rivers is the state of the oxygen regime. If levels of oxygen are too low, fish and plant life dies. Dead rivers are those that contain no oxygen, while we mostly refer to low oxygen levels when we talk about polluted streams. Oxygen levels in rivers are influenced by the amount of dead organic matter present and the consequent bacterial activity, as the bacteria responsible for decay of dead organic matter in water require oxygen to do their job. If there is too much bacterial activity, insufficient oxygen might be left for the fish life. The sources of dead organic matter, in turn, are primarily waste water treatment plants as well as a range of natural sources such a dead leaves, agricultural runoff (accounts for approximately 33% of dead organic matter present in rivers), and urban sewage. All streams possess a natural cleansing ability allowing them to degrade organic waste. Problems arise when the stream is overloaded with the latter, and its ecosystem breaks down. A second cause of changes in oxygen levels are the nutrients contained in runoff from fertilized agricultural areas; the excess algae growth induced by higher levels of nitrogen and phosphorus likewise competes with fish life and organisms for oxygen.

Two indicators of the state of the oxygen regime are employed, as different stations and countries report one or the other, but not always both. Dissolved oxygen indicates the level of oxygen in a river. A more common measure for water quality, BOD, is an inverse measure of the levels of dissolved oxygen. It reports the amount of oxygen required for biochemical decomposition. Previous research studying the relationship between the state of the oxygen regime and per capita income levels has arrived at conflicting results. Grossman and Krueger report an environmental Kuznets curve for both DO_2 and BOD, while Shafik and Bandyopadhyay found DO_2 to increase with per capita income levels.

[2] Because of the skewed distribution of SPM levels, the log of the dependent variable is utilized to normalize the data.

The analysis concentrates on measurements of river water quality, as data on lake and groundwater stations is too sparse. For both DO_2 and BOD, the mean levels are documented, ($\mu g/m^3$). Importantly, the majority of stations reports the mean temperature of the river, which has a significant influence on oxygen levels, and needs to be included in the analysis as a control variable. Data on BOD were available for 215 stations in 39 countries and DO_2 for 263 stations in 50 countries between 1979 and 1990.

Fecal coliform is a measure used to indicate the disease potential of a water resource caused by pathogenic pollution. Compared to the state of the oxygen regime, pathogenic pollution has more to do with human access to safe drinking water than with the health of aquatic resources. The lack of clean, disease-free drinking water is still one of the most serious water pollution problems today. While proper treatment of water has greatly reduced the risk of epidemics in most developed countries, every year billions of people, especially in developing countries, face the risk of waterborne diseases. Pathogens from fecal discharges can cause a range of diseases ranging from an upset stomach to potentially fatal diseases such as typhoid and cholera. Specifically, fecal coliform counts measure microbial pollution, mostly by common and harmless forms of bacteria. The presence of these harmless bacteria, however, indicates the likelihood of the presence of the harmful bacteria, which are difficult to measure directly. The sources of pathogenic pollution are all kinds of human and animal waste. Separating sewage from drinking water supplies is, thus, an essential step in reducing the disease potential of water resources. Furthermore, in most developing countries the water supply is continuously being monitored for quality.

Previous research again found conflicting results when studying the relationship between fecal coliform and per capita income levels. Grossman and Krueger report GDP per capita to be associated with constant levels of fecal coliform up to a per capita income of $8,000 and to fall sharply thereafter. Shafik and Bandyopadhyay, however, find a cubic relationship.

Coliform is measured as log(1+Y), where Y is number of organisms per 100 milliliters. The transformation is necessary because of the exponential growth in coliform and the skewed nature of the data. A simple log operation is not possible, since many of the stations report coliform levels of zero when the reading falls below the detectable level of the measuring instrument. Furthermore, a control variable for the mean annual water temperature is included since fecal coliform is especially sensitive to heat, so that bacterial counts can vary over several orders of magnitude at a given monitoring station, as well as a control variable for the type of measuring device used. The sample includes measurements from 205 stations in 44 countries between 1979 and 1990.

Deforestation

Deforestation is probably one of the most visible environmental problems today. Even though this visibility is deceptive in terms of the identification of the 'villains,' since many nations cut down their forests over the past centuries and therefore have low deforestation rates today, the problem demands attention. Deforestation is such a serious problem because of the historical link between forests and civilization and because of the multiple functions forests fulfil. Forests are economic resources providing wood for building material, as a source of fuel, as material for boats and wagons, apart from any symbolic, aesthetic and recreational functions. Nearly half of the world's population still depends on wood for cooking as well as heating. Moreover, forests reduce problems of soil erosion, act as water sheds, and provide a habitat for wildlife including many endangered species. The irreversibility of deforestation is a further problem since genetic diversity is lost which diminishes future opportunities to develop pharmaceuticals (a very anthropocentric concern). Most importantly, forests are pivotal for climate protection as they absorb atmospheric carbon dioxide, a greenhouse gas. Naturally, deforestation reduces this 'storage space.' In addition, deforestation often takes place in the form of burning which releases the carbon sequestered in the forest's biomass and thereby compounds the problem.

Deforestation is measured as change in forest cover in nations over time. Forests are defined as areas where trees are the dominant vegetation ranging from closed canopy forests to open woodlands, often also scrublands and bushlands. While changes in forest cover are natural and necessary, this natural change is extremely slow in terms of human life expectancy. Today forests are cut at alarmingly high rates for lumber and paper products, for fuel, or to obtain agricultural land. Only approximately 2.5 billion hectares (ha) of closed forest and 1.2 billion ha of open woodlands and savannas remain in the world. Furthermore, indirect deforestation, the death of trees from pollution, is a reality but not counted in the deforestation data.

As pointed out earlier, several studies of smaller cross-sections of countries have been able to find support for a significant relationship between property rights and deforestation. Interestingly, the relationship between per capita income levels and environmental quality, so prevalent for all other environmental indicators, could not be proven by Shafik and Bandyopadhyay (1992). The authors explain this lack of evidence with the 'bias' inherent in the sample, as we lack data on the historical cutting of forests by the now-developed nations.

Deforestation data, specifically data on the annual percentage of deforestation, are gathered by the World Resources Institute. Unfortunately, the official description of the data, as providing data on deforestation rates in a cross-section of countries from 1980-1990, is misleading, as only one data point in the form of the average annual deforestation rate and the average percentage change in forest cover for the entire decade are reported for each country. The sample therefore is small, including only 89 observations (from 89 countries).

Access to Safe Drinking Water and Sanitation

Access to safe drinking water and sanitation are important determinants of the quality of the environment in which we are living. For more than a billion people worldwide the lack of access to safe drinking water and sanitation pose serious health problems, including a generally high morbidity and mortality from waterborne diseases. Recognizing this, countries set targets for the degree of access to drinking water and sanitation to be achieved by 1990. At that point, approximately 1.3 billion people still lacked access to safe drinking water, while 1.7 billion lacked adequate sanitation. Access to water is a major component necessary for sustainable development. Yet resources are becoming ever more scarce, with more people and functions (power companies, agricultural and leisure uses) competing for them. As a consequence, experts are questioning whether access to drinking water and sanitation is actually increasing or decreasing in developing countries. Previous research, however, has found it to increase with increases in per capita income levels, which would support the intuitive logic that governments will provide and people will obtain these services once they can afford them (Shafik and Bandyopadhyay, 1992).

Data on access to safe drinking water and sanitation were obtained from the World Bank's Economic and Social Indicators data base as well as the WRI. Both sources report percentages of the population with access to these services, separated into rural and urban. All four measures are used as dependent variables in the analysis: percent of urban population with access to safe drinking water, percent of rural population with access to safe drinking water, percent of urban population with access to sanitation, and percent of rural population with access to sanitation.[3] The sample includes observations from 108 countries for access to drinking water, and observations from 69 countries for access to sanitation.

[3] As with most indicators of environmental quality, the data on access to safe drinking water and sanitation are far from perfect. Access to safe water is not a question of being connected to a supply system, but also of cost and quality of water provision. Measuring coverage as the percentage of population having access to standpipes or house connections is therefore an inaccurate representation of whether people have access to safe water. (World Bank, 1995: 33f) Furthermore, besides the general paucity of data, definitional uncertainties regarding 'adequate amount' or 'safe' make comparisons over time difficult. At the same time, private or self-provision of these basic services, which people will rely on if government provision is not available but private resources are is often not counted.

What Is Missing

Important indicators of environmental quality are absent from this analysis. They include other measures of urban air pollution such as nitrogen and carbon monoxides or lead, alternative measures of water pollution such as contamination by heavy metals, and, most importantly, indicators of other dimensions of environmental quality such as soil degradation or measures of biodiversity. Their absence is particularly unfortunate, since they include environmental indicators for which the distancing between the consumption and production of environmental degradation has generally confounded scholars pursuing this type of empirical analyses. Institutional capacity in the United States, for instance, might do little to protect biodiversity in the Amazon region (or in fact might do the opposite). Likewise, carbon monoxide emissions are likely to be less effected by institutional capacity as the consequences are global rather than local in nature. Supportive results of the environmental impact of institutional capacity, then, might overestimate their effect to some degree or simply cannot be generalized to all dimensions of environmental quality.

A.2 The Independent Variables

Institutional Capacity

The central focus of this book is on the environmental implications of institutions, in particular property rights. Therefore, the empirical analysis was supposed to include a measure of property rights. Clague, Keefer, Knack, and Olson (1996, 1995) developed such a measure in the form of 'Contract Intensive Money' (CIM) to estimate the impact of secure property rights on economic growth. The measure is based on the idea that individuals in environments with insecure property rights will choose to engage only in self-enforcing contracts, which provide less potential for the society to realize gains from trade and build a foundation for economic growth. Contracts requiring external enforcement by a third party will only be chosen if the institutional capacities exist to provide this enforcement. Consequently, the designers of this measure argue that the relative reliance on non-currency monetary instruments should mirror the perceived security of property rights.

Clague et al. introduce the CIM measure when analyzing the differentials in levels and rates of growth and investment between countries and trying to explain why the poor countries do not catch up. They hypothesize that the poor performance of some countries is caused by a lack of institutional arrangements necessary to reap the larger gains from trade, which arise from enforcement-intensive contracts. Contracts requiring external enforcement by a third party will only be chosen if the institutional capacities exist to provide this enforcement. While markets for self-enforcing contracts can be found in any economy, the authors point out that most contractual arrangements in sophisticated economies depend on the expectation that an authoritarian body will enforce the contract if necessary. Consequently, the authors expect the ratio of 'contract-intensive, institution-sensitive transactions' to 'spot-market and self-enforcing transactions' to be higher in economically successful societies with capital-intensive production than in poor societies with labor-intensive production. Furthermore, they argue that this ratio should be related to economic growth and rates of capital investment.

Operationalizing this difference between self-enforcing contracts and those requiring intensive external contract enforcement, Clague et al. argue that the sophistication of the financial instruments used in a society will mirror the conditions of contract enforcement. If property rights are insecure, loans to institutions or people will be insecure as well, and people will prefer cash (or even barter) exchanges. Based on this argument, then, Clague et al. derive their measure of contract-intensive money.

> Thus the extent to which societies can capture not only the gains from self-enforcing transactions, but also those potential trades that are intensive in contract enforcement and property rights, can be approximated by the relative use of currency in comparison with "contract-intensive money." We define contract-intensive money (CIM) as the ratio of non-currency money to the total money supply $[M_2\text{-}C]/M_2$, where M_2 is a broad definition of the money supply and C is currency held outside banks (p. 7).

With this measure, the authors test their hypothesis that the CIM ratio of a country is positively related to the transaction-friendliness of its institutions, gains from trade and specialization reaped, investment, per capita incomes, and economic growth. Their empirical results show that a higher CIM ratio means indeed higher per capita incomes, economic growth and investment.

Clague et al. argue that the CIM measure offers a new indicator of the quality of institutions. Some scholars will look at this measure critically, however, and argue that all it captures are levels of development, especially in terms of development of financial markets. The authors show both logically and empirically that this is not the case. They test the significance of CIM when controlling for GDP per capita, financial sophistication (measured as M_2/GDP), price stability, and inflation, and find that the results are robust to the addition of these variables. Furthermore, CIM performs well for both OECD and non-OECD countries, indicating that it does not simply capture the differences between developed and developing nations.[4]

The CIM measure, thus, was considered as a monetary approximation of the security of property rights for this analysis. However, it appears that CIM measures general institutional capacity of a government rather than property rights specific aspects, as the creators of the measure claim. After all, the general institutional capacity of a government is an important determinant of the development of sophisticated financial markets and institutions. These markets and institutions need to be based on prudent and effective regulation in order to be efficient and trustworthy. Thus, in this analysis, CIM is utilized to measure the influence of governmental institutional capacity on environmental quality. The data for CIM, in particular on M_2 and the share of currency in M_2, were obtained from the International Financial Statistics (IFS) of the International Monetary Fund (IMF), where they are available for a large number of countries over a significant period of time.

Democracy

Political scientists have identified numerous political factors as determinants of policy outcomes, and specifically environmental quality. As pointed out in chapter 2, one of the most prominent arguments regarding the political determinants of environmental quality concerns regime types and centralization. Accordingly, both factors have been included in the empirical analysis.

Are democracies better or worse protectors of the environment? Both sides of this debate have logical arguments in their favor. One the one side, democracies might be incapable of solving today's major environmental problems, as the cause of these problems is just the liberty awarded to self-interested individuals to pursue their personal gain without concern for the social benefit. Individual rationality might prevent the necessary sacrifice and the solution of the associated collective action problems (Hardin, 1968; Heilbroner, 1975; Ophuls, 1977). On the other side, a higher responsiveness to public demands for a clean environment, higher degree of long-term legitimacy awarded to environmental policies, improved information flows, and a consequently more knowledgeable public might make democracies superior in solving environmental problems (Press, 1994; Passmore, 1974; Payne, 1995). Moreover, the openness of the system, which gives access to business and labor interests and therefore rise to criticism of the environmental potential of democracies, may also give access to environmental organizations. Likewise, the central role of the market system

[4] Again, these results were robust to additions of measures of inflation, financial depth, population growth, trade openness, government size, and increases in human capital.

in Western democracies may provide a better foundation green consumerism and marketing.

The divergence in views arises from differing assessments of the individual's ability to decipher complex ecological problems and pursue solutions through effective participation in public affairs. Advocates of the 'authoritarian' side do not have to look far to find evidence that this ability is limited. Democracies have indeed proven unable to deal with some of the most severe environmental problems to date, and the influence of special interests on representatives is well documented. Furthermore, the long-term legitimacy allegedly awarded to environmental policies by the democratic process is not necessarily given, as the current political dynamics in the United States show. The question is, however, whether non-democratic regimes have, on average, produced superior results because of their ability to overcome the democratic collective action problems. Increasingly, better information about the state of the environment in Eastern Europe seems to suggest that that is not the case. The former socialist regimes developed environmental collective action problems of their own and left, in some instances, abhorrent environmental conditions behind. Furthermore, advocates of the 'democracy benefits environment' argument point out that environmental organizations were among the first to develop in the transition to democracy in many Eastern European states.[5]

Again, the debate on the environmental impacts of differences in levels of centralization in the decision-making process is closely associated with the above one. *Centralists* argue that centralized decision-making is better for environmental quality because it is associated with a higher potential to achieve the optimal social outcome (Heilbroner, 1975; Ophuls, 1977; McIntyre and Thornton, 1978). They base their claims on the view that a centralized decision-making process reduces the coordination difficulties between the various units, and allows more efficient gathering and distribution of information. Most importantly, centralized policies avoid competition between the various units through top-down channels. In the context of environmental standards, for instance, one concern with policy making at the state or even local level is that states or communities will try to attract business by down-competing environmental standards.

Decentralists, in contrast, argue that decision-making at lower levels and by more units is better for the environment (Press, 1994; Passmore, 1974;

[5] This argument has to be taken with a grain of salt, however, since many environmental organizations developed before the transition and received part of their popularity from their unofficial status and opposition role to the socialist regime and government. It has yet to be shown how they will fare under the democratic regime in the long term, especially given the pressing economic problems, which reduce the significance of environmental problems in the minds of many people (as we know all too well from the experiences of Western environmental groups).

Schumacher, 1989; Goldsmith, 1972; Lovins, 1977). They highlight the capabilities of a decentralized systems to adjust to local needs and allow a higher level of participation. Furthermore, they blame centralization as part of the problem, as it leads to production processes and technologies beyond the control of the individual, the consequences and costs of which are rarely understood: "(...) the crisis owes in part to the loss of a sense of appropriate scale and meaningful purpose" (Orr and Hill, 1978: 465).

Similar to the democracy debate, a number of contributions to the decentralization argument have favored middle-of-the-road solutions, pointing out, for example, that the appropriate level of centralization is a function of the environmental issue at hand. Orr and Hill (1978), for instance, argue for selective decentralization with centralized coordination as the best political structure to promote a 'resilient society' (see also Orr, 1992). The main criticism that *centralists* have to face is that they assume a centralized state can cope with its own increased size and complexity, need for skill, exertion of control, and still be efficient at the management of a diverse range of tasks. Such assumptions are increasingly viewed with doubt both by scholars and practitioners, and - similar to the democracy side - *decentralists* are currently holding the upper hand in the debate. *Decentralists*, however, have yet to come up with answers to criticisms they face, such as how to avoid coordination difficulties and most of all how to ensure that competition between units will not lead to weaker environmental measures. The federal model, then, could be seen as a way to combine the advantages of centralization and decentralization, and allow the necessary flexibility in the choice of the appropriate policy strategy.

One of the greatest difficulties when it comes to empirical tests of the impact of political factors is their operationalization. How do you measure democracy? Fortunately, several data sets measuring political characteristics and recording political events have been developed in the last decades. For this specific test, data from Gurr's Polity III data set were employed. Polity III contains data on authority traits for a large number of countries between 1800 and 1987. In contrast to Gastil's Civil and Political Liberties Index, which Gastil codes on the basis of what the substance of rights seems to be in countries according to media reports, Gurr's measures of democracy and others are based on the legal provisions given in each country.

In order to assess the claims of the positive or negative influences of democracy and centralization on environmental quality, the two respective measures from the data set are employed. The democracy indicator combines ratings in four categories to compute a weighted ten point scale: competitiveness in political participation, competitiveness of executive recruitment, openness of executive recruitment, and constraints on the chief executive. Gurr bases this choice of categories on grounds that they are 'essential, interdependent elements' in contrast to 'means to, or specific manifestations of theses general principles,' such as a system of checks and balances or freedom of the press (Gurr, Jaggers, and Moore, 1991: 79).

Environmental Effort

Finally, the analysis assesses the influence of the environmental effort of government on environmental quality. While this factors was not discussed as a determinant of environmental quality that is central in the literature in chapter 2, the analyses of chapters 4 and 5 have demonstrated that a government's commitment to environmental objectives is an important determinant of environmental outcomes. In a related inquiry, this empirical analysis attempts to assess to what extent empirical evidence supports the claim that a government's environmental effort has an influence on the environmental quality achieved at a given level of development. Intuition suggests and environmental policy assumes that the environmental policy effort of a government would have a positive impact on environmental quality in a nation.

As with the political indicators discussed above, however, we return to the problem of operationalization when we want to empirically test this intuitively convincing argument. How can we compare environmental effort on a cross-national basis? While there are several possibilities to approximate governmental environmental effort, for instance in the form of environmental budgets or the signing of international environmental agreements, each of these has significant flaws. Looking at budgets does not tell us how efficiently the money is spent, or, more importantly, how much is lost to corruption. Moreover, for many nations we would have to combine federal and state if not local budgets, as environmental policies are often decided and implemented at the state or local levels, which again leads to difficulties in terms of data availability and comparability. International agreements, on the other side, are often signed for publicity reasons, while their provisions are not implemented and enforced, and therefore fail to provide a trustworthy picture as well.

For this preliminary analysis, then, date were obtained from the World Bank's coding of national environmental reports. These reports were prepared by 145 countries for the United Nations Conference on Environment and Development (UNCED, 1992). Their contents is somewhat comparable because the UN had prescribed a standard format. Furthermore, while admittedly the information in the reports is based on self-reporting, the requirement that NGOs and business representatives be part of the drafting process seems to have assured to some degree that these reports are more than governmental propaganda: "To a striking degree, they seem to reflect real environmental conditions and issues" (Dasgupta, Moody, Roy and Wheeler, 1995: 3).

Based on these reports, Dasgupta et al. (1995) developed a set of comparative indices for the status of environmental policy and performance in the respective countries. They randomly selected 31 reports from the 145 provided to UNCED, with countries ranging from highly industrialized to extremely poor, and representing every region in the world. The authors considered the status of policy and performance in four environmental dimensions, air, water, land, and living resources, and attempted to assess the interaction between each of these and five 'active categories': Agriculture, Industry, Energy, Transport, and the Urban Sector. In order to arrive at a quantitative measure of the status and performance of environmental policy, they applied a survey with 25 questions to categorize (a) the state of environmental awareness; (b) scope of policies adopted; (c) scope of legislation enacted; (d) control mechanisms in place; and (v) the degree of success in implementation, giving 0-2 points for low, medium and high rankings. In sum, the coding provided 500 assessment scores answering each of the 25 questions for each of the 20 elements in the matrix for each UNCED report (see table A1).

Table A1. Evaluation Format Used by Dasgupta et al.

Sector/Activity	Air	Water	Land	Living Resources
Agriculture				
Industry				
Energy				
Transport				
Urban				

From the individual scores, the authors calculated four composite indices by adding scores within each dimension and also computed a total score for the state of environmental policy. For the empirical analysis conducted here, the significance of both the composite score in the respective dimension of the environmental indicator and the total score was tested. Given the small number of countries for which I have these data, however, the indicator could only be utilized for regressions on one measure of air quality (SO_2) and the three indicators of water quality.

The Control Variables

In the choice of control variables, the analysis is guided by previous research. Since Grossman and Krueger (1995, 1993), Jänicke et al. (1995), and Shafik and Bandyopadhyay (1992) all found GDP per capita to be strongly related to environmental quality, the analysis controls for its influence by including the necessary GDP terms in the equation.

Furthermore, a time trend is included to reduce error from attributing improvements in environmental quality that derive from technological innovation to institutional capacity or any other of the factors under scrutiny. Finally, the analysis utilizes a measure of the mean temperature of the water in the regressions on the indicators of water quality, since temperature has a strong impact on fecal coliform counts, as well as biochemical oxygen demand and dissolved oxygen levels.[6]

On the one hand, it seems that there is nothing much to say about GDP per capita as a control variable, since it is such a common economic indicator. The GDP per capita data used here come from the Chain Index World Penn Tables by Summers and Heston and provide measures on per capita output in relation to a common set of international prices. On the other hand, however, books have been written about the deficiencies of per capita GDP as a measure of development. It fails, for instance, to include the informal sector or household economy and counts indicators of social problems as assets. Alternative indices including literacy rates or infant mortality have been constructed because of these deficiencies. Most importantly, should not environmental disasters, the loss of natural capital, and environmental defensive expenditures, which positively contribute to the conventional GDP measure, be deducted from our measure of per capita incomes, especially in the context of this research project? Substantial amounts of research are being pursued on such a greening of national income measures. Unfortunately, however, cross-national data for the 1970s and 1980s are not available yet. Moreover, such a change in indicators would inhibit this effort to replicate and extend the previous research on the relationship between per capita income levels and environmental quality. Therefore, establishing the true relationship between environmental quality and green per capita income levels will remain a task for future research.

In the regression analyses, up to six forms of GDP per capita are included, corresponding to the findings Grossman and Krueger, Shafik and Bandyopadhyay, and Jänicke et al. that the relationship between GDP per capita and environmental quality is non-linear and depends on previous income levels as well. To capture the influence of these previous income levels, Grossman and Krueger designed a measure of 'permanent income' calculated as the average per capita income of the three previous years, which this analysis uses as well. The six GDP per capita terms are: GDP per capita, squared GDP per capita, cubed GDP per capita, permanent GDP per capita, squared permanent GDP per capita, and cubed permanent GDP per capita.

[6] Grossman and Krueger use a larger set of control variables. Since this analysis concentrates on the 'within' variation, however, control variables which are constant for a specific cross-section such as locational data do not need to be included.

A.3 The Econometric Technique

The choice of estimation technique is primarily determined by the characteristics of the data in the sample, and by the assumptions we made about the relationship between the sample and the population. In this case, several of the environmental indicators allowed a panel estimation because observations cover measuring stations and countries over time. Panel data allow for different possibilities regarding varying intercepts and slopes across time and cross-sectional units. Among these models, one of the most common is based on constant slope coefficients and varying intercepts over cross-sectional units. The strength of this model is that it captures effects of omitted variables that are individual, time-invariant, i.e. constant for a given cross-sectional unit through time but varying across cross-sectional units. The variable-intercept model assumes that the effects of the omitted time-invariant individual variables "can be absorbed into the intercept term of a regression model as a means to explicitly allow for the individual (...) heterogeneity contained in the temporal cross-sectional data" (Hsiang, 1986: 26). By introducing these unit-specific variables into our panel specification, then, we are able to reduce or avoid the omitted-variable bias. In addition, there are different specifications from which to choose. The basic decision is whether to apply a fixed-effects or random-effects model to the estimation. Unfortunately, the theoretical justifications for this choice are not well established and quite controversial among econometricians. Moreover, previous research cannot offer much guidance in this choice, since Shafik and Bandyopadhyay utilized a fixed-effects model, while Grossman and Krueger chose a random-effects one. This analysis utilizes the fixed-effects model, since the arguments in its favor are more convincing.[7] Many of the

[7] Mundlak (1978) criticized the random-effects model for cases in which omitted variables are correlated with regressors, under which circumstances we would arrive at biased estimations. This is probably the case in these samples as the clustering of developed countries in some regions of the world and of developing countries in others causes environmental spill-overs. A fixed-effects model concentrating on within-country variance, allows, for instance, to reduce the errors from attributing SO_2 emissions from Germany or the European continent in general to Sweden, or from attributing pollution of the Ganges caused by India to Bangladesh. Moreover, according to Hsiang (1986), we need to be able to assume that the cross-sectional units in our sample do indeed form a random sample of members of the larger population in order to use the random-effects model, which then allows us to make general inferences from this sample to the population. While drawing inferences from the countries in our sample to the population of countries would be ideal, we cannot claim that the countries represented in our data form a random sample of the entire population of countries in the world. In most of the panels, OECD countries are over-represented, for instance, while they are underrepresented in a few others, most notably access to safe water and sanitation. Greene (1993) presents a counter argument, however. He postulates that the use of a fixed-effects

country-specific determinants of environmental quality omitted from the model, such as the geographic conditions or population size, will correlate with some of the independent variables. Consequently, the time-invariant, cross-section specific intercepts in the fixed-effects model allow to reduce the bias resulting from the omission of these variables. In addition, transboundary pollution, which would cause more of a problem in a random-effects model than in a fixed-effects one, is indeed a serious problem when we look at measures of water quality in rivers, and measures of air quality. Furthermore, as pointed out above, given the prevalence of OECD countries in some of the samples (and their under-representation in a few others), it would be difficult to argue that the sample is based on a random drawing from the population of countries. Unfortunately, the choice of the fixed-effects model meant a substantial loss in efficiency of the estimates, however, because of the relatively large size of N compared to T. The coefficients are estimated using the Feasible Generalized Least Squares Estimator (FGLS) with robust standard errors based on a variance/covariance matrix of errors calculated according to Huber's (or White's) method.

In those cases in which the time series in the sample were too short, a meaningful estimation of "within"-variation was not possible. This applies to the indicators of access to safe drinking water and sanitation. In the corresponding regressions, therefore, the random effects estimator was used. Fortunately, the theoretical considerations that spoke against using a random-effects model for the indicators of air and river quality do not weigh as heavily here. Transboundary effects and environmental spillovers are not as influential in determining the level of access to safe drinking water and sanitation, as they are for explaining air and water pollution. While some influence might derive from trade or other interdependencies between countries, it will be much less significant than for levels of air and water quality.

Finally, in the case of deforestation the Ordinary Least Squares (OLS) estimator is used to estimate the regression. Given the absence of heteroskedasticity and serial correlation, the OLS estimator is BLUE. Fortunately, almost none of the samples showed significant serial correlation.[8]

---

model is appropriate in cases where the sample is exhaustive, i.e. includes the full set of countries. In contrast, he claims that it would be more appropriate to use the random-effects model and "to view individual specific intercepts as randomly distributed across cross-sectional units if the sample is drawn from a large population" (Greene, 1993: 469).

[8] The empirical analysis was further complicated, however, by the presence of strong multicollinearity in the samples. Given the presence of six GDP per capita terms this was not surprising. Unfortunately, there is no easy way of correcting for multicollinearity unless it is possible to drop the highly-correlated variables or replace them with

Using the institutional capacity equation as an example, the estimated model then is as follows:

$$EQ_{it} = \alpha_i + CIM_{it}\beta_1 + (GDP/cap)_{it}\beta_2 + (GDP/cap)^2_{it}\beta_3 + (GDP/cap)^3_{it}\beta_4 + (LGDP/cap)_{it}\beta_5 + (LGDP/cap)^2_{it}\beta_6 + (LGDP/cap)^3_{it}\beta_7 + Time_t\beta_8 + C_{it}\beta_9 + \varepsilon_{it}$$

where EQ is a vector of the respective indicator of environmental quality, α is a vector of country specific intercepts, Time is a linear time trend, C is a vector of covariates (which are used for determining the characteristics of the data, such as heteroskedasticity, but dropped for the within-estimation), and ε_{it} is the error term.

A.4 The Evidence

Institutional Capacity

The results of the empirical tests of the influence of institutional capacity (CIM) on environmental quality generally indicated a positive relationship, as the analyses in chapters 4 and 5 would have suggested. The relationship between the measure of institutional capacity and environmental quality follows the expected form and is significant for eight of the eleven indicators of environmental quality. Institutional capacity is positively related to access to rural and urban sanitation and safe drinking water, as well as oxygen levels in rivers, and negatively related to most forms of environmental pollution tested: smoke, SO_2, and BOD. Most importantly, in most of these

---

instruments that capture the intended aspect but are less strongly correlated with the other regressors (a concept which can only work in the rarest of cases). In other cases, conducting a log operation on the variables in the analysis will reduce the collinearity. Obviously, I did not have an instrument which I could substitute for the various GDP terms, and did not want to drop the terms permanently, since they had been found to be significant by Grossman and Krueger, and Shafik and Bandyopadhyay. Furthermore, a log operation did not work in this case, since taking a log on the squared and cubed GDP terms would only have turned the high collinearity into perfect collinearity. Grossman and Krueger dealt with the multicollinearity by avoiding interpreting the individual coefficients estimated, and instead tested the joint significance of the GDP terms. Similarly, I handled the problems in two ways. First, I estimated each regression with the institutional capacity measure and the time trend and one of the GDP terms at a time. This gave me an idea of which of the GDP terms was more significant while at the same time providing me with an estimate of the significance of the institutional capacity variable, which is not that strongly correlated with the GDP terms as to render my results invalid. Secondly, I estimated each panel with all of the GDP terms as well as the CIM measure and the time trend, after which I tested the significance of the CIM measure and the joint significance of the GDP terms individually and jointly, in order to assess whether the CIM measure added to the explanatory power of the model.

regressions the institutional capacity variable is as significant or even more significant than the GDP terms.

Sulfur Dioxide

In the case of SO_2, the coefficient on the CIM variable is significant and as pointed out above by being negative has the expected sign (see Table A2). In other words, improvements in the institutional capacity of government mean lower levels of SO_2. The coefficient is not as significant as the GDP terms and does not increase the joint significance of those terms. With an R^2 of .77, the model also explains a substantial share of the variance in SO_2 levels in the sample. SO_2 emissions from power plants and other industrial processes, thus appear to decrease with the increasing institutional capacity of government. With respect to the other independent variables, the GDP terms are strongly significant, in the individual regressions as well as jointly in the complete model. The coefficient on the time trend has a negative sign and is marginally significant, indicating that at any given income level SO_2 levels have on average been decreasing over the years.

Smoke

Similarly, the relationship between institutional capacity and environmental quality takes on the expected form in the case of smoke suggesting that emissions of light particulate matter decrease with improvements in institutional capacity. The coefficient on the CIM variable is negative, implying that a better institutional capacity means lower levels of smoke (see Table A3). Furthermore, the coefficient is strongly significant, and improves the joint significance of the GDP terms when added to the test. With an R^2 of .83 (adjusted R^2 .79), the model again explains a large share of the variance in levels of smoke in the sample. In contrast to the CIM measure, the GDP terms are not significant in the individual regressions and only marginally significant in the complete model. Furthermore, the time trend is positively related to level of smoke, i.e. at any given level of institutional capacity and per capita incomes, smoke has increased during the time range in the sample.

Table A2. Results for SO2

$SO_2 = \alpha + \beta_1 CIM + \beta_2 X + \beta_3 TIME$				Number of obs: 1304 R-squared: .77 Adj. R-squared: .72
Regression	Regressor	Coefficient (st.error)	T-statistic	p-value
X=GDP/cap	CIM	-60.38 (28.12)	-2.15	.032
	GDP/cap	-5.91 (1.14)	-5.17	.000
X=(GDP/cap)^2	CIM	-70.17 (31.01)	-2.26	.024
	(GDP/cap)^2	-.193 (.046)	-4.15	.000
X=(GDP/cap)^3	CIM	63.54 (32.7)	1.94	.052
	(GDP/cap)^3	-.006 (.002)	-2.97	.003
X=(LGDP/cap)	CIM	-56.4 (27.71)	-2.035	.042
	(LGDP/cap)	-8.14 (1.32)	-6.181	.000
X=(LGDP/cap)^2	CIM	-74.72 (29.52)	-2.53	.011
	(LGDP/cap)^2	-.318 (.06)	-5.298	.000
X=(LGDP/cap)^3	CIM	-69.43 (31.55)	-2.2	.028
	(LGDP/cap)^3	-.012 (.003)	-3.831	.000

complete model:

$SO_2 = \alpha + \beta_1 CIM + \beta_2 GDP/cap + \beta_3 (GDP/cap)^2 + \beta_4 (GDP/cap)^3 + \beta_5 (LGDP/cap) + b_6 (LGDP/cap)^2 + \beta_7 (LGDP/cap)^3 + \beta_8 TIME$

coefficient (standard error)	t-statistic	R-squared (adjusted R-squared)
- 67.65 (29.91)	2.262	.78 (.74)
significance of CIM (p-value)	joint significance - GDP terms (p-value)	joint significance CIM, GDP terms (p-value)
5.12 (.024)	15.96 (.000)	14.35 (.000)

Table A3. Results for Smoke

$SMOKE = \alpha + \beta_1 CIM + \beta_2 X + \beta_3 TIME$				Number of obs: 463 R-squared: .83 Adj. R-squared: .79
Regression	Regressor	Coefficient (st.error)	T-statistic	p-value
X=GDP/cap	CIM	-57.12 (32.88)	1.74	.083
	GDP/cap	-.744 (1.66)	0.45	.654
$X=(GDP/cap)^2$	CIM	-58.52 (31.67)	1.85	.065
	$(GDP/cap)^2$	-.05 (.08)	.62	.539
$X=(GDP/cap)^3$	CIM	-58.8 (31.56)	1.86	.063
	$(GDP/cap)^3$	-.003 (.005)	0.64	.520
X=(LGDP/cap)	CIM	-64.2 (33.15)	1.94	.054
	(LGDP/cap)	.203 (1.89)	1.07	.285
$X=(LGDP/cap)^2$	CIM	-57.99 (32.36)	1.79	.074
	$(LGDP/cap)^2$	.006 (.113)	0.05	.960
$X=(LGDP/cap)^3$	CIM	-57.95 (31.95)	1.81	.070
	$(LGDP/cap)^3$	-.006 (.008)	.757	.450

complete model:
$Smoke = \alpha + \beta_1 CIM + \beta_2 GDP/cap + \beta_3 (GDP/cap)^2 + \beta_4 (GDP/cap)^3 + \beta_5 (LGDP/cap) + b_6 (LGDP/cap)^2 + \beta_7 (LGDP/cap)^3 + \beta_8 TIME$

coefficient (standard error)	t-statistic	R-squared (adjusted R-squared)
- 117.96 (35.75)	3.30	.83 (.79)
significance of CIM (p-value)	joint significance - GDP terms (p-value)	joint significance CIM, GDP terms (p-value)
10.89 (.001)	2.27 (.036)	2.95 (.005)

Table A4. Results for Heavy Particulates

$SPM = \alpha + \beta_1 CIM + \beta_2 X + \beta_3 TIME$				Number of obs: 839 R-squared: .93 Adj. R-squared: .91
Regression	Regressor	Coefficient (st.error)	T-statistic	p-value
X=GDP/cap	CIM	1.06 (.44)	2.384	.017
	GDP/cap	.01 (.01)	.507	.612
X=$(GDP/cap)^2$	CIM	.98 (.47)	2.105	.036
	$(GDP/cap)^2$	-.0004 (.0005)	.767	.443
X=$(GDP/cap)^3$	CIM	.91 (.47)	1.926	.054
	$(GDP/cap)^3$	-.00003 (.00002)	1.399	.162
X=(LGDP/cap)	CIM	1.04 (.45)	2.329	.020
	(LGDP/cap)	-.002 (.02)	.156	.876
X=$(LGDP/cap)^2$	CIM	.92 (.47)	1.965	.050
	$(LGDP/cap)^2$	-.001 (.0007)	1.488	.137
X=$(LGDP/cap)^3$	CIM	.84 (.48)	1.761	.079
	$(LGDP/cap)^3$	-.00007 (.00003)	2.160	.031

complete model:
$SPM = \alpha + \beta_1 CIM + \beta_2 GDP/cap + \beta_3 (GDP/cap)^2 + \beta_4 (GDP/cap)^3 + \beta_5 (LGDP/cap) + b_6 (LGDP/cap)^2 + \beta_7 (LGDP/cap)^3 + \beta_8 TIME$

coefficient (standard error)	t-statistic	R-squared (adjusted R-squared)
97.71 (102.77)	.951	.93 (.92)
significance of CIM (p-value)	joint significance - GDP terms (p-value)	joint significance CIM, GDP terms (p-value)
.45 (.502)	3.58 (.002)	4.10 (.000)

Heavy Particulates

For heavy particulate matter the coefficient on the CIM measure has a positive sign, however, suggesting a positive association between institutional capacity and pollution (see Table A4). It is only marginally significant in the individual regressions, and adds only slightly to the joint significance of the GDP terms in the complete model, but that does not change the results regarding the form of the relationship. One can therefore only posit that assurance of institutional capacity positively influences some sources of particulate matter but not others. The GDP terms, at the same time, are significant in the complete model, while the coefficient on the time trend is negative as in the case of SO_2. Over the years, pollution by heavy particulate matter has improved at given levels of institutional capacity and per capita incomes.

Biochemical Oxygen Demand

For BOD, the coefficient on the institutional capacity measure has the expected negative sign attesting that an improvement in institutional capacity will lead to lower levels of BOD (see Table A5). Furthermore, the coefficient is statistically significant, in fact more significant than the individual or joint GDP terms, and adds substantially to their joint significance in the complete model. The results suggest, then, that water pollution leading to increasing consumption of oxygen falls with increasing institutional capacity. Again, with an R^2 of .74 (adjusted R^2 of .68), the model explains a substantial share of the variance of BOD levels in the sample. In terms of the other independent variables, I find that mean temperature is positively related to BOD, which makes intuitive sense, but not significant. Similarly, the coefficient on the time trend is positive, i.e. BOD levels have risen at any given level of institutional capacity and income over the period in the sample, but is not significant.

Dissolved Oxygen

Again, the positive coefficient on the CIM measure reflects the improvement in DO_2 that is associated with improvements in institutional capacity (see Table A6).

Table A5. Results for BOD

$BOD = \alpha + \beta_1 CIM + \beta_2 X + \beta_3 TIME + \beta_4 Meantemp$				Number of obs: 1207 R-squared: .74 Adj. R-squared: .68
Regression	Regressor	Coefficient (st.error)	T-statistic	p-value
X=GDP/cap	CIM	-19.08 (7.86)	2.43	.015
	GDP/cap	-.223 (.25)	0.89	.373
$X=(GDP/cap)^2$	CIM	-19.24 (8.00)	2.41	.016
	$(GDP/cap)^2$	-.015 (.009)	1.69	.092
$X=(GDP/cap)^3$	CIM	-19.44 (8.06)	2.41	.016
	$(GDP/cap)^3$	-.0008 (.0004)	1.97	.049
X=(LGDP/cap)	CIM	-18.98 (7.94)	2.39	.017
	(LGDP/cap)	-.504 (.282)	1.79	.074
$X=(LGDP/cap)^2$	CIM	-19.35 (8.04)	2.41	.016
	$(LGDP/cap)^2$	-.023 (.011)	1.99	.047
$X=(LGDP/cap)^3$	CIM	-19.51 (8.08)	2.41	.016
	$(LGDP/cap)^3$	-.001 (.0007)	2.00	.046

complete model:
$BOD = \alpha + \beta_1 CIM + \beta_2 GDP/cap + \beta_3 (GDP/cap)^2 + \beta_4 (GDP/cap)^3 + \beta_5 (LGDP/cap) + b_6 (LGDP/cap)^2 + \beta_7 (LGDP/cap)^3 + \beta_8 TIME$

coefficient (standard error)	t-statistic	R-squared (adjusted R-squared)
-20.68 (7.89)	2.621	.74 .68
significance of CIM (p-value)	joint significance - GDP terms (p-value)	joint significance CIM, GDP terms (p-value)
6.87 (.009)	1.77 (.102)	1.82 (.080)

Table A6. Results for DO_2

$DO_2 = \alpha + \beta_1 CIM + \beta_2 X + \beta_3 TIME + \beta_4 Meantemp$				Number of obs: 1529 R-squared: .86 Adj. R-squared: .83
Regression	Regressor	Coefficient (st.error)	T-statistic	p-value
X=GDP/cap	CIM	3.39 (1.58)	2.15	.032
	GDP/cap	.176 (.053)	3.3	.001
$X=(GDP/cap)^2$	CIM	3.58 (1.55)	2.45	.015
	$(GDP/cap)^2$	.009 (.002)	4.252	.000
$X=(GDP/cap)^3$	CIM	3.78 (1.54)	2.31	.021
	$(GDP/cap)^3$	.0005 (.0001)	4.118	.000
X=(LGDP/cap)	CIM	3.54 (1.55)	2.29	0.22
	(LGDP/cap)	.249 (.068)	3.68	.000
$X=(LGDP/cap)^2$	CIM	3.66 (1.55)	2.37	.018
	$(LGDP/cap)^2$	.015 (.003)	4.57	.000
$X=(LGDP/cap)^3$	CIM	3.79 (1.55)	2.46	.014
	$(LGDP/cap)^3$	.0009 (.0002)	4.704	.000

complete model:
$DO_2 = \alpha + \beta_1 CIM + \beta_2 GDP/cap + \beta_3 (GDP/cap)^2 + \beta_4 (GDP/cap)^3 + \beta_5 (LGDP/cap) + b_6 (LGDP/cap)^2 + \beta_7 (LGDP/cap)^3 + \beta_8 TIME + \beta_9 Meantemp$

coefficient (standard error)	t-statistic	R-squared (adjusted R-squared)
4.76 (1.66)	2.871	.86 (.83)
significance of CIM (p-value)	joint significance - GDP terms (p-value)	joint significance CIM, GDP terms (p-value)
8.24 (.004)	5.09 (.000)	5.31 (.000)

Table A7. Results for Fecal Coliform

$\log(1+FEC) = \alpha + \beta_1 CIM + \beta_2 X + \beta_3 TIME + \beta_4 Meantemp$				Number of obs: 1216 R-squared: .82 Adj. R-squared: .78
Regression	Regressor	Coefficient (st.error)	T-statistic	p-value
X=GDP/cap	CIM	10.14 (2.89)	3.511	.000
	GDP/cap	-.153 (.085)0	1.92	.055
X=(GDP/cap)^2	CIM	9.77 (2.86)	3.42	.001
	(GDP/cap)^2	-.005 (.003)	1.51	.132
X=(GDP/cap)^3	CIM	9.66 (2.86)	3.38	.0001
	(GDP/cap)^3	-.0003 (.0002)	1.67	.095
X=(LGDP/cap)	CIM	10.12 (2.86)	3.54	.000
	(LGDP/cap)	-.256 (.111)	2.30	.022
X=(LGDP/cap)^2	CIM	9.88 (2.85)	3.46	.001
	(LGDP/cap)^2	-.0098 (.0048)	2.06	.040
X=(LGDP/cap)^3	CIM	9.77 (2.85)	3.42	.001
	(LGDP/cap)^3	-.0015 (.0002)	2.11	.035

complete model:
$\log(1+FEC) = \alpha + \beta_1 CIM + \beta_2 GDP/cap + \beta_3 (GDP/cap)^2 + \beta_4 (GDP/cap)^3 + \beta_5 (LGDP/cap) + b_6 (LGDP/cap)^2 + \beta_7 (LGDP/cap)^3\ \beta_8 TIME + \beta_9 Meantemp$

coefficient (standard error)	t-statistic	R-squared (adjusted R-squared)
10.51 (2.91)	3.613	.82 (.78)
significance of CIM (p-value)	joint significance - GDP terms (p-value)	joint significance CIM, GDP terms (p-value)
13.06 (.0003)	7.46 (.000)	8.78 (.000)

The relationship is statistically significant, although it only slightly improves the joint significance of the GDP terms when added in the test (the GDP terms are strongly significant). The R^2 of .86 (adjusted R^2 .83) further supports the strength of the model. Similar to the results for BOD, the results for DO_2, then, suggest that water pollution, resulting in higher oxygen consumption by bacteria and excess algae growth and lower levels of dissolved oxygen, falls with improvements in institutional capacity. Again, the negative coefficient on mean water temperatures also delineates the expected relationship; as water temperature goes up, oxygen levels go down. This dynamic can easily be explained with the biochemical effects of increases in water temperature: higher temperatures mean more algae and bacterial growth and generally a lower capability of the water to hold oxygen. The coefficient on the time trend is negative as well reflecting an overall worsening of DO_2 levels.

Fecal Coliform

Fecal coliform counts also form an indicator of water quality in rivers for which improvements in institutional capacity appear to lead to improvements in environmental quality (see Table A7). The sign of the coefficient for institutional capacity is negative, suggesting that environmental quality improves with improvements in institutional capacity. The results suggest, then, that water pollution from the untreated discharge of human and animal waste decreases with improvements in institutional capacity. The coefficient is significant at the 0.5 level, but does not add to the joint significance of the GDP terms in the complete model. The GDP terms individually fail to show strong significance, but are jointly significant in the complete model. Again, mean water temperature is positively related to pollution levels, as could be expected, but not significant. Finally, the coefficient on the time trend is positive and strongly significant denoting an overall worsening of water quality in terms of fecal coliform counts at any given level of institutional capacity or income over the period of the sample.

Access to Safe Drinking Water[9]

The results for urban and rural access to safe drinking water indicate a positive relationship between institutional capacity and environmental quality (see Tables A8 and A9). The coefficient on the CIM measure is positive and significant. Improvements in institutional capacity lead to access to safe drinking water for larger shares of both urban and rural populations. Interestingly, institutional capacity appears to have slightly more explanatory power for urban rather than rural conditions. Furthermore,

[9] As pointed out above, the results reported here pertain to the estimation of the random effects model. For the results of the fixed-effects estimations see Appendix B.

the CIM variable is statistically as significant as the GDP terms, and substantially improves their joint significance in the complete model. As could be expected, the time trend is positively related to access to drinking water implying an improvement in conditions at any given level of institutional capacity and incomes over time. Interestingly, it is much more strongly significant in the regression for rural than for urban access.

Access to Sanitation

Again, for both urban and rural access to sanitation, the regression estimates show a positive relationship between the CIM measure and environmental quality (see Tables A10 and A11). Moreover, the CIM measure is statistically significant in the individual regressions as well as in the complete model. In addition, institutional capacity substantially improves the joint significance of the explanatory variables. The time trend is also positively related to urban and rural access to sanitation, although for rural access it is not significant. Finally the GDP terms are significant, with lagged per capita GDP performing best in the individual regressions, implying maybe a delay in the response between rising income levels and the provision of sanitation services.

Deforestation

In the case of deforestation, the regression results show no evidence that institutional capacity is positively associated with environmental quality as indicated by decreases in deforestation (see Table A12). Not only is the coefficient for institutional capacity not significant, but it is positive, and does not add to the joint significance of the GDP terms in the model. The R^2 is correspondingly low, with the model explaining only .23 of the variance in annual percentages of deforestation in the sample. The GDP terms, in contrast, are significant individually as well as jointly in the complete model, and a time trend, of course, was not included, since the data contained only one observation per country.

Table A8. Results for Rural Access to Safe Drinking Water

model: random effects Rural access to Safe Drinking Water = β_1CIM + β_2X + β_3TIME				Number of obs: 289 R-squared: .70
Regression	Regressor	Coefficient (st.error)	T-statistic	p-value
X=GDP/cap	CIM	25.65 (13.05)	1.966	.049
	GDP/cap	3.89 (.424)	9.175	.000
X=$(GDP/cap)^2$	CIM	54.71 (13.61)	4.019	.000
	$(GDP/cap)^2$	.091 (.020)	4.539	.000
X=$(GDP/cap)^3$	CIM	63.20 (13.81)	4.577	.000
	$(GDP/cap)^3$	.001 (.0007)	1.820	.069
X=(LGDP/cap)	CIM	16.15 (12.68)	1.273	.203
	(LGDP/cap)	4.90 (.462)	10.601	.000
X=$(LGDP/cap)^2$	CIM	38.34 (13.12)	2.923	.003
	$(LGDP/cap)^2$	.228 (.030)	7.628	.000
X=$(LGDP/cap)^3$	CIM	51.46 (13.45)	3.825	.000
	$(LGDP/cap)^3$	.009 (.002)	5.284	.000

complete model:
rur.wat. = β_1CIM + β_2GDP/cap + $\beta_3(GDP/cap)^2$ + $\beta_4(GDP/cap)^3$ + β_5(LGDP/cap) + $b_6(LGDP/cap)^2$ + $\beta_7(LGDP/cap)^3$ + β_8TIME

coefficient (standard error)	t-statistic	R-squared within R-squared between
2.56 (13.08)	.211	.28 .70
significance of CIM (p-value)	joint significance - GDP terms (p-value)	joint significance CIM, GDP terms (p-value)
.04 (.833)	142.35 (.000)	196.99 (.000)

Table A9. Results for Urban Access to Safe Drinking Water

model: random effects Urban access to Safe Drinking Water = β_1CIM + β_2X + β_3TIME				Number of obs: 290 R-squared: .51
Regression	Regressor	Coefficient (st.error)	T-statistic	p-value
X=GDP/cap	CIM	49.76 (10.97)	4.535	.000
	GDP/cap	1.23 (.355)	3.453	.001
X=$(GDP/cap)^2$	CIM	60.24 (10.51)	5.729	.000
	$(GDP/cap)^2$	.027 (.015)	1.723	.085
X=$(GDP/cap)^3$	CIM	63.59 (10.31)	6.166	.000
	$(GDP/cap)^3$	.0004 (.0005)	.779	.436
X=(LGDP/cap)	CIM	46.49 (10.98)	4.235	.000
	(LGDP/cap)	1.48 (.400)	3.703	.000
X=$(LGDP/cap)^2$	CIM	54.88 (10.67)	5.142	.000
	$(LGDP/cap)^2$	.059 (.024)	2.442	.015
X=$(LGDP/cap)^3$	CIM	58.94 (10.46)	5.633	.000
	$(LGDP/cap)^3$	.002 (.001)	1.686	.092

complete model:
urb.wat. = β_1CIM + β_2GDP/cap + $\beta_3(GDP/cap)^2$ + $\beta_4(GDP/cap)^3$ + β_5(LGDP/cap) + $b_6(LGDP/cap)^2$ + $\beta_7(LGDP/cap)^3$ + β_8TIME

coefficient (standard error)	t-statistic	R-squared within R-squared between
34.15 (11.37)	3.005	.06 .51
significance of CIM (p-value)	joint significance - GDP terms (p-value)	joint significance CIM, GDP terms (p-value)
9.03 (.003)	31.87 (.000)	82.55 (.000)

Table A10. Results for Rural Access to Sanitation

model: random effects — Number of obs: 99

Rural access to Sanitation = β_1CIM + β_2X + β_3TIME — R-squared: .32

Regression	Regressor	Coefficient (st.error)	T-statistic	p-value
X=GDP/cap	CIM	65.31 (25.89)	2.523	.012
	GDP/cap	.630 (.847)	.743	.457
X=(GDP/cap)^2	CIM	73.12 (25.18)	2.904	.004
	(GDP/cap)^2	-.010 (.028)	.340	.734
X=(GDP/cap)^3	CIM	74.63 (24.97)	2.989	.003
	(GDP/cap)^3	-.0006 (.0009)	.646	.518
X=(LGDP/cap)	CIM	54.42 (24.99)	2.177	.029
	(LGDP/cap)	4.23 (1.51)	2.795	.005
X=(LGDP/cap)^2	CIM	65.26 (24.69)	2.644	.008
	(LGDP/cap)^2	.298 (.139)	2.145	.032
X=(LGDP/cap)^3	CIM	70.05 (24.68)	2.838	.005
	(LGDP/cap)^3	.021 (.012)	1.710	.087

complete model:

rur.san. = β_1CIM + β_2GDP/cap + β_3(GDP/cap)^2 + β_4(GDP/cap)^3 + β_5(LGDP/cap) + b_6(LGDP/cap)^2 + β_7(LGDP/cap)^3 + β_8TIME

coefficient (standard error)	t-statistic	R-squared within R-squared between
50.39 (26.37)	1.911	.14 .32
significance of CIM (p-value)	joint significance - GDP terms (p-value)	joint significance CIM, GDP terms (p-value)
3.65 (.056)	13.47 (.036)	23.56 (.001)

Table A11. Results for Urban Access to Sanitation

model: random effects Urban access to Sanitation = β_1CIM + β_2X + β_3TIME				Number of obs: 111 R-squared: .35
Regression	Regressor	Coefficient (st.error)	T-statistic	p-value
X=GDP/cap	CIM	60.77 (22.17)	2.741	.006
	GDP/cap	1.72 (.72)	2.390	.017
X=$(GDP/cap)^2$	CIM	72.18 (21.90)	3.295	.001
	$(GDP/cap)^2$	.031 (.025)	1.232	.218
X=$(GDP/cap)^3$	CIM	59.29 (21.06)	2.815	.006
	$(GDP/cap)^3$	.001 (.001)	.948	.345
X=(LGDP/cap)	CIM	50.06 (22.26)	2.249	.025
	(LGDP/cap)	4.23 (1.27)	3.345	.001
X=$(LGDP/cap)^2$	CIM	64.60 (22.03)	2.933	.003
	$(LGDP/cap)^2$	.269 (.115)	2.337	.019
X=$(LGDP/cap)^3$	CIM	74.18 (21.86)	3.393	.001
	$(LGDP/cap)^3$	.0007 (.0008)	.898	.369

complete model:
urb.san. = β_1CIM + β_2GDP/cap + $\beta_3(GDP/cap)^2$ + $\beta_4(GDP/cap)^3$ + β_5(LGDP/cap) + $b_6(LGDP/cap)^2$ + $\beta_7(LGDP/cap)^3$ + β_8TIME

coefficient (standard error)	t-statistic	R-squared within R-squared between
33.6 (23.83)	1.410	.18 .35
significance of CIM (p-value)	joint significance - GDP terms (p-value)	joint significance CIM, GDP terms (p-value)
1.99 (.159)	18.43 (.005)	31.27 (.000)

Table A12. Results for Deforestation

deforestation = $\alpha + \beta_1 CIM + \beta_2 X + \beta_3 TIME$				Number of obs: 82 R-squared: .23
Regression	Regressor	Coefficient (st.error)	T-statistic	p-value
X=GDP/cap	CIM	1.83 (.97)	1.885	.063
	GDP/cap	-.15 (.03)	4.689	.000
X=$(GDP/cap)^2$	CIM	1.07 (.93)	1.149	.254
	$(GDP/cap)^2$	-.01 (.002)	4.145	.000
X=$(GDP/cap)^3$	CIM	.56 (.92)	.605	.547
	$(GDP/cap)^3$	-.001 (.0001)	3.489	.001
X=(LGDP/cap)	CIM	1.82 (.98)	1.863	.066
	(LGDP/cap)	-.16 (.03)	4.608	.000
X=$(LGDP/cap)^2$	CIM	1.09 (.93)	1.164	.248
	$(LGDP/cap)^2$	-.01 (.002)	4.115	.000
X=$(LGDP/cap)^3$	CIM	.59 (.92)	.641	.523
	$(LGDP/cap)^3$	-.001 (.0002)	3.505	.0001

complete model:
deforestation = $\alpha + \beta_1 CIM + \beta_2 GDP/cap + \beta_3 (GDP/cap)^2 + \beta_4 (GDP/cap)^3 + \beta_5 (LGDP/cap) + b_6 (LGDP/cap)^2 + \beta_7 (LGDP/cap)^3 + \beta_8 TIME$

coefficient (standard error)	t-statistic	R-squared (adjusted R-squared)
1.91 (.93)	2.049	.28 (.21)
significance of CIM (p-value)	joint significance - GDP terms (p-value)	joint significance CIM, GDP terms (p-value)
1.64 (.201)	2.74 (.004)	2.43 (.007)

Democracy and Centralization

The empirical analyses of the influence of democracy and centralization on environmental quality show why the different parties in the democracy and centralization debates all have been able to find some evidence for their claims. At the same time, they highlight that maybe none of these claims can be empirically supported across a range of environmental dimensions and cases. For both regime type and levels of centralization, the influence on environmental quality can be positive or negative depending on the indicator. More importantly, in most cases neither of the indicators is statistically significant. In a few cases, however, the indicators are significant, in some even more significant than institutional capacity. In sum, we find an erratic picture that brings up more questions than it answers.

Table A13 Results for Democracy and Centralization

Environ. Indicator	Democracy	p-value	adds to joint significance	Centrali-zation	p-value	adds to joint significance
SO_2	2.54	.005	./.	-9.07	.004	./.
SPM	-.025	.096	+	-.064	.289	+
smoke	1.06	.040	./.	-11.23	.0003	./.
BOD	-.037	.696	+	-.485	.454	+
DO_2	-.017	.357	./.	.333	.259	./.
fec.coliform	-.109	.023	./.	-.421	.298	./.
rural water	-.056	.898	+	-4.21	.147	+
urban water	.402	.250	+	1.06	.652	+
rural sanit.	-1.31	.205	./.	-3.17	.578	./.
urban sanit.	.189	.823	./.	-5.61	.263	./.
deforestation	.072	.181	+	-.465	.033	+

The democracy indicator is individually significant only for SO_2, fecal coliform levels, and smoke. In the cases of access to water, SPM, BOD, and deforestation, it adds to the joint significance of the explanatory variables, which highlights multicollinearity problems in the data. While the indicator is more significant than institutional capacity in the case of SPM, this finding is not surprising given the bad performance of institutional capacity in this specific instance. The relationship between democracy and environmental quality according to the empirical findings is slightly more positive (for urban access to safe drinking water and sanitation, DO_2, SO_2, smoke, and deforestation) than negative (for rural access to drinking water and sanitation, SPM, BOD, and fecal coliform) with the statistically most significant coefficient (SO_2) being positive, and the statistically second most significant coefficient (fecal coliform) being negative. As pointed out above, we find an erratic pattern that casts doubt even on the significance of (the measured characteristics of) government regimes for the few indicators of environmental quality for which the results identified a significant causal relationship.

A similar failure to convincingly identify a statistically significant influence on environmental quality applies to the indicator of centralization. While the coefficient on centralization is strongly significant for smoke and SO_2, it is not significant for the majority of indicators. According to the sign on the coefficient in the different regressions, the relationship between levels of centralization and environmental quality is mostly negative, implying that decentralization is associated with improved environmental quality. This relationship can be identified for rural access to drinking water, access to sanitation, SO_2, SPM smoke, BOD, DO_2, and fecal coliform. The only case in which decentralization appears to lead to a worsening of environmental quality is urban access to safe drinking water. As pointed out above, however, the vast majority of these relationships is not statistically significant, so that we cannot place much confidence in these interpretations. Yet importantly, the relationship is negative for the statistically significant cases SO_2 and smoke, implying that the decentralists might have an edge over the centralists.

One hope for participants in these debates arising from the rather ambiguous and disappointing results, however, is that the addition of indicators of regime types and levels of centralization significantly improves the R^2 in many of the regressions. Together, then, these indicators hold some explanatory power. An implication of the findings, therefore, might be that we need to move to more differentiated and sophisticated analyses of the interactive dynamics between regime types, levels of centralization, and environmental quality. Thus, one might first have to look at the distribution of cost and benefits of the actions necessary for environmental improvement, as well as the associated transaction cost in order to answer the question whether a democratic or an authoritarian regime is better for environmental quality. If the costs for most members of society are significant either financially or in terms of requiring a change in deeply ingrained habits, an authoritarian government might accomplish an environmental policy objective faster than a democratic one. On the other side, if the costs of environmental actions fall only on a few big manufacturers, for instance, and public pressure is sufficiently high, one might achieve more momentum in a democratic political system than from a government structure not as responsive to public demands.

Environmental Effort

The empirical evidence for the influence of governmental environmental effort on environmental quality is ambiguous at best. The indicators of environmental effort are individually significant only in the cases of SO_2 and urban access to safe drinking water. The high correlation between the indices and per capita GDP, however, might blur the underlying relationships. In spite of the lack of individual significance, the overall environmental effort measure does add substantially to the significance of the complete model in five of the nine regressions (four of the five regressions for the dimension-specific ones respectively). Again, the small number of observations in most of the samples requires us to be careful when drawing conclusions from these results.

Table A14. Results for Governmental Environmental Effort

Environ. Indicator	air/water/ liv. resources	p-value	adds to joint signif.	environmental effort	p-value	adds to joint signif.
SO_2	2.69	.000	+	.785	.000	+
BOD	.054	.532	./.	.011	.63	+
DO_2	-.002	.897	+	-.002	.62	+
fec.coliform	-.059	.094	+	-.017	.093	./.
rural water	n/a	n/a	n/a	-.051	.586	./.
urban water	n/a	n/a	n/a	.227	.007	+
rural sanit.	n/a	n/a	n/a	.068	.797	./.
urban sanit.	n/a	n/a	n/a	.218	.181	./.
deforestation	.0008	.960	+	.0009	.801	+

The sign of the coefficient of the effort measure is counter-intuitive in many of the regressions, indicating that a higher score in terms of effort is associated with lower environmental quality. This finding suggests that the relationship we are capturing might have the wrong causal direction: environmental effort may be high in cases where environmental quality is bad, and government intervention is especially necessary. Unfortunately, cross-sections, in contrast to data sets including a time series aspect, do not allow to test for temporal sequence.

A.5. Discussion

Due to the conflicting results on the influence of democracy and centralization on environmental quality, and the potentially simple explanation of the counter-intuitive findings on the influence of environmental effort, this discussion will concentrate on the empirical results with respect to institutional capacity. Overall, the empirical results are supportive of the positive influence of institutional capacity on environmental quality, both in terms of direction and significance. For the majority of indicators, the empirical evidence strongly indicates that institutional capacity leads to improvements in environmental quality. While per capita income levels are generally also statistically significant - thus attesting to the overall congruence of the results with the findings of previous research - in most cases institutional capacity is more strongly related to environmental quality. This result is fascinating, since previous research had failed to find any political or economic factor to have a stronger influence on environmental quality than per capita income levels. Moreover, cases in which the results fail to show a positive influence of institutional capacity can be explained, if we consider interactive dynamics between capacity, development, and the specific indicator of environmental quality. Only the case of fecal coliform confounds the analysis, if we remind ourselves that the cross-sectional analysis of deforestation is not commensurate with the analyses of the variation within cross-sectional units of the other environmental indicators. In sum, these findings show that improvements in institutional capacity can be linked to improvements in environmental quality in terms of air and water quality, as well as access to drinking water and sanitation.

Interestingly, the evidence is particularly strong for the environmental quality of traditional open-access resources such as air and river water. The cases of SO_2, smoke, BOD, and DO_2, all show that improvements in institutional capacity translate into improvements in environmental quality. This pattern may be a function of regulatory capacity with respect to the environment. It may also be caused by "private" or political action, or economic incentives. One possible incentive can derive from public pressure in terms of consumer demand. Thus, chemical companies along the Rhine valley had to invest substantial resources into cleaner technology after several accidents lead to highly visible consequences in the form of massive fish dying, to avoid further public scrutiny. Pressure for improvements in environmental quality which are based on individual incentives might also be channelled through the political process, and result in the imposition of more stringent environmental regulation or the establishment of private incentives for polluters.

The counter case, i.e. the failure of institutional capacity to lead to improvements in environmental quality in the case of heavy particulate matter, opens a range of interesting questions, as well. The positive sign of the institutional capacity coefficient can suggest the necessity to take a closer look at the interactive dynamics in the model. One reason, why the sign is 'wrong' may be that better institutional capacity will generally induce more development. Increased construction, which is a consequence of development, is also one of the major sources of heavy particulate matter. Therefore, the positive association between heavy particulate matter and a high level of institutional capacity may derive from the interaction between the causal factors. A naive person might argue that we therefore should prefer less institutional capacity in order to avoid development. The alternative, however, no development or improvement in living standards, is difficult to sell to Americans or inhabitants of other developed countries. It is certainly not an option for developing countries. Rather than avoiding development, we need to focus on sustainable development, and I would argue that institutional capacity is an important element here.

In addition, the stronger significance of institutional capacity for the share of the urban populations with access to safe drinking water, versus rural ones, illustrates an interesting difference between these two indicators, which on the surface would appear to be almost identical. When taken together with the fact that the time trend is stronger for rural access, this finding suggests that urban access to safe drinking water is comparatively more driven by the immediate politico-economic conditions, i.e. institutional capacity and financial resources and independent of time. Rural access, in contrast appears cross-nationally to have strongly improved with time reflecting maybe a concerted political effort as discussed above, given the absence of sufficiently favorable economic conditions in rural areas at that point.

Compared to the supportive results for access to sanitation and drinking water and most of the indicators of air and water quality, the results for deforestation are disappointing. Of course, the failure to find the expected relationship may be a function of the quality of the data. In the case of deforestation, the sample is extremely small and the estimates are of notably poor quality.[10]

In general, however, the results strongly suggest the potential of institutional capacity to benefit environmental quality in various dimensions of the latter. The implication, then, is that governments can provide a supportive institutional framework for improvements in environmental quality in conjunction with development.

A closer look at the other explanatory variables provides further interesting insights. In terms of the time trend, for instance, we would like to know whether we generally see an improvement in environmental quality at any given level of institutional capacity and incomes over time, which would indicate a learning curve in the form of technological innovation. In fact, Grossman and Krueger introduced the time trend specifically for the purpose of capturing the effect of improvements in technology, in order to not attribute them to changes in per capita incomes. The results, however, tell a different story. In five of the ten regressions that have a time trend, this time trend is negatively related to environmental quality. BOD, smoke, DO_2, and fecal coliform, all have deteriorated over time at any given level of income and institutional capacity. Interestingly, this group contains the three indicators of water quality, but only one indicator of air quality, which may be pointing to the greater improvements in cleaning technology or just

[10] Shafik and Bandyopadhyay highlight the paucity of deforestation data in their study. Their empirical results for deforestation are similarly weak.

stronger efforts in clean air policy than in clean water policy. Alternatively, population growth and the continuously increasing use of fertilizers worldwide may be causing the negative trend for water quality, as these factors impact the selected indicators of water quality more directly than those of air quality. Improvements over time have come for access to drinking water and sanitation, where we know that concerted political efforts have been made, as well as for SO_2 and SPM. As pointed out earlier, SO_2 is a cause of acid rain and has, in this context, received political attention in many developed countries as well. The positive developments for these indicators independent of income levels underline what is also illustrated by the institutional capacity measure: where there is a will there is a way.

Rather than pushing these findings too strongly, however, consider an important aspect of the results that should cause some concern. A significant need for further research in this area becomes clear when we compare the present findings on the influence of per capita income levels on environmental quality to those of the previous research efforts. On the positive side, some of the results are in agreement with the findings by either Grossman and Krueger, Shafik and Bandyopadhyay, or Jänicke and his colleagues. On the negative side, however, they agree sometimes with one research team and at other points with another, and sometimes even with none of the previous findings at all.

Access to safe drinking water is a case with little controversy. Similar to Shafik and Bandyopadhyay, the present analysis finds access to uniformly increase with higher per capita income levels. In the case of SPM, however, the finding of an inverted u-curve corresponds to Shafik and Bandyopadhyay's finding, but contradicts the findings by both Grossman and Krueger as well as Jänicke et al., who find particulates to uniformly decrease with per capita income levels. In contrast, in the case of DO_2, the present finding of an inverted u-curve corresponds to Grossman and Krueger's results, but contradicts Shafik and Bandyopadhyay's finding of a uniform worsening of oxygen levels with rising per capita incomes. Significantly, in both cases, the turning points in the calculations here are much higher than those calculated by Grossman and Krueger and Shafik and Bandyopadhyay respectively. Finally, in the case of fecal coliform, the present findings agree with neither those of Grossman and Krueger, nor those of Shafik and Bandyopadhyay. In spite of the generally supportive results, then, what becomes obvious is how little we really know about the complex dynamics in our ecological system. While some of these differences in findings might arise from different compositions in the samples, they generally highlight how much more we need to scrutinize theoretically and empirically the relationships between the different causal factors.

An additional caveat to the predictive strength of these empirical assessments is that the results describe the relationships of the past, particularly of the seventies and eighties. There are no guarantees that the relationships found will be the same in the 21st century or even in the nineties. Probably the relationships between environmental quality and the different causal factors were different before the seventies and eighties as well, especially before the rise of the environmental movement. The reason for the changing relationships is that they depend on given economic and especially technological conditions at a time, as well as social conditions specifically environmental awareness. Improvements in technology, for instance, can mean that at any given income level or for any level of institutional capacity, countries will pollute less. Countries that are currently industrializing or will do so in the 21st century might not have to go through the SO_2 phase. Likewise, international pressure for environmental improvement or even domestic pressures arising as knowledge about environment related health concerns travels around the world can influence these relationships. As a consequence of such developments, we might find higher environmental quality at any given level of institutional capacity and per capita incomes if not a different form of relationship in the future.

At the same time, however, we might also witness the opposite, i.e. that improvements in per capita incomes or even institutional capacity might not be associated with parallel improvements in environmental quality. In the past, developed countries have often achieved improvements of environmental quality, especially air quality, not only through the imposition of stringent air quality regulations, but also through the movement of the most polluting industries to developing countries. That means, however, that some of the developing countries will not have the same option to improve their air quality.

In sum, the empirical evidence supports the expected positive influence of institutional capacity on environmental quality. At any level of development, improvements in institutional capacity lead to improvements in environmental quality in terms of air and water quality, as well as in terms of access to safe drinking water and sanitation for both rural and urban populations. Future research will need to analyze the environmental implications of institutional capacity for additional indicators of environmental quality, as well as attempt to solve the puzzle of conflicting results for the relationship between per capita income levels and environmental quality.

Appendix B

In addition to the institutional factors discussed in Appendix A, preliminary assessments of the empirical support for social and cultural factors as determinants of environmental quality were conducted. The respective factors tested were levels of post-materialism in a society and income distributions.

B.1 Culture

The identification of culture as a determinant of environmental policy derives its main impetus from the work by Inglehart (1971, 1990), and is based on the argument that changes in the culture of Western, industrialized societies have led to corresponding changes in the kind of economic growth these societies are pursuing (see also Inglehart and Abramson, 1994). According to Inglehart, these changes, originally caused by developments in the economic, political, and technological conditions of society, have led to an increased concern for quality of life aspects, including environmental quality, and a lessened focus on economic growth rates.

As pointed out above, empirical research has found post-materialism to be a strong determinant of environmental values (Dalton, 1994; Müller-Rommel, 1990; Rohrschneider, 1993). Arguments by other scholars, however, link environmental values to higher incomes, education and the influence of younger generations, thus focusing on social rather than culturally specific conditions.

Data to empirically test the support for Inglehart's argument were obtained from the ICPSR world values survey.[1] In this data set, two variables represent a post-materialism index. The variables are constructed on the basis of 4 and 12 items, respectively, which in turn were derived from answers to 2 and 3 questions pertaining to the interviewee's most important personal goals and political concerns. Since for the first of these two variables, the indicator constructed on the basis of four items, data are available for more countries in my sample, the analysis was forced to rely on that one. Furthermore, data on this indicators are available for a number of countries for 1971 and for a slightly smaller group of countries for 1981. While further away from our sample data in temporal terms, the analysis utilized the 1971 data in this preliminary analysis, again simply because it provided a larger sample, and the opportunity to arrive at results that have some statistical validity. Already, the limited number of countries in Inglehart's data set meant that the analysis could empirically assess the relationship between post-materialism and environmental quality only for four indicators of air and water quality, and deforestation, if it was to be based on a sufficient sample size. Furthermore, the analysis had to rely on cross-sectional analyses instead of the previously used panel estimations for the indicators of air and water quality (the same applies to the estimation of the influence of income variance below).

As the results of the empirical analysis show, the post-materialism indicator performs slightly better than those of government regimes and centralization discussed in Appendix A The relationship between post-materialism and environmental quality follows the expected form and is significant in four of the six regressions tested, i.e. slightly more than 50% of the cases. In the cases of SO_2 and BOD, the post-materialism index individually performs even better than institutional capacity. Still, the indicator adds to the joint significance of the complete model only in the cases of SO_2 and DO_2, and substantially improves the R^2 of the model only for SO_2.

Table B1. Results for Culture

Environmental Indicator	Culture	p-value	adds to joint significance
SO2	-46.93	.000	+
smoke	126.29	.013	./.
BOD	8.89	.059	./.
DO2	-1.74	.003	./.
fec. coliform	1.22	.000	./.
deforestation	.533	.217	./.

[1] Since the data-set is based on surveys, the data are on an individual level, of course. In order to achieve congruence with my other data in terms of level of analysis, the analysis is based on the national means as derived from the survey data.

The sign of the post-materialism coefficient also tells a more unambiguous story than in the cases of democracy and centralization. For the statistically significant relationships between post-materialism and environmental quality, the coefficient is generally positive. Yet, in one of the cases with the highest significance of the indicators, SO_2, it is negative, pointing to higher levels of SO_2 in cultures with higher levels of post-materialism. With this surprising result, the lack of substantial improvements of the R^2 and the joint significance of the complete model, then, the empirical support for the importance of post-materialism for environmental quality is not very convincing as well. The very small sizes of some of the samples in these estimations, of course, are partly to blame for this. The significance of the post-materialism index in most equations suggests the benefit of further exploration of the interactive dynamics between cultural characteristics and environmental quality.

B.2 Income Distribution

Finally, what impact do differences in income distributions have on the environmental quality achieved at any given level of development, if per capita income is such a strong determinant of environmental quality? Going beyond the fundamentally flawed measure of individual welfare provided by per capita GDP, the actual income of the majority of the population should have a significant impact on the use of natural resources. Including a measure of income distributions in the empirical assessment still does not allow an exact measure of individual welfare, but improves on the assessment of a population's actual wealth and income provided by per capita GDP.

A highly uneven distribution of wealth, for instance, could mean that the majority of a population depends on the (over)exploitation of natural resources for every-day survival. Furthermore, without the means to protect resources, their decision-making will be similar to that of appropriators from open-access resources. Consequently, environmental problems are likely to be distributed in a geographically and socially highly uneven fashion. While the elite is likely to be enjoying relatively healthy environmental conditions with functioning sewage systems and access to safe drinking water, and may have a concern for clean air, clean rivers, and beautiful landscapes in the proximity of their homes, the poor sections of the population lack the financial means to improve their respective environmental conditions.

Even the natural resources owned by the wealthy elite, however, may not fare well under these circumstances, if we look at the political consequences of skewed income distributions. If the income is distributed highly unevenly in a country, the political climate, and with it government's institutional capacity, might be unstable, as a large group of dissatisfied people always holds potential for an upheaval. As discussed earlier, an elite which finds itself in the lucky situation of having achieved ownership over natural resources might decide to exploit these as fast as possible, before its luck changes.

Again, data on income distributions are not easy to find, especially when looking for time series data. For this preliminary assessment, the analysis relied on cross-sectional data provided by the World Bank and Keefer and Knack (1994). From the World Bank's Economic and Social Indicators Database, the analysis obtained data on the share of national income held by the upper 20 and lower 40 percent of the population, from which an income distribution ratio was calculated. Keefer and Knack found this ratio to perform equivalent to or in some cases even better than the GINI coefficient, when looking at the relationship between inequality and economic growth for a cross-section of countries.

The empirical results for the influence of income variance on environmental quality, unfortunately, are as ambiguous as those of some of the indicators discussed above. The coefficients on the ratio terms are significant in about one third of the regressions, although they are strongly significant only for fecal coliform, DO_2, and SO_2. Generally, the ratio variable performs better than the GINI coefficient, which is strongly significant only in the case of DO_2, and significant in the cases of SO_2, smoke, and fecal coliform. In the cases of SO_2, smoke, and fecal coliform, the ratio is individually more significant than the CIM measure, but it only adds to the joint significance of the complete model in the case of DO_2. Moreover, in all of the regressions the improvements in the adjusted R^2 are small.

Table B2. Results for Income Variance

Environ. Indicator	Ratio	p-value	adds to joint significance	GINI coeff.	p-value	adds to joint significance
SO_2	16.0	.000	./.	1.07	.016	./.
SPM	1.27	.066	./.	.030	.207	+
smoke	7.84	.375	./.	.876	.048	./.
BOD	.219	.829	./.	-.334	.299	./.
DO_2	.312	.003	+	.055	.002	+
fec.coliform	-.501	.008	./.	-.072	.019	./.
rural water	-2.18	.385	./.	-.276	.573	./.
urban water	-1.25	.505	./.	-.162	.660	./.
rural sanit.	-3.90	.216	+./.	-.874	.120	./.
urban sanit.	-4.51	.118	+./.	-.592	.318	./.
deforestation	-.018	.837	+	-.010	.496	+

What story does the sign on the coefficients of the ratio and GINI terms tell us? Again, the coefficients indicate a positive relationship between environmental quality and income distributions in some cases and a negative one in others, with the majority, however, suggesting that environmental quality is poorer where income distributions are highly skewed. The picture is less clear, however, if we restrict our inquiry to the significant relationships. For two of the four indicators for which income distribution is significant, we find a positive relationship between income distributions and environmental quality. Oxygen levels and fecal coliform counts improve the more the income distribution is skewed to the wealthiest quintile of the population. In contrast, SPM and SO_2 levels worsen with higher ratios. Again, the overall pattern is rather erratic offering little conclusive information. Contrary to intuition, we cannot confidently claim that income variances influence environmental quality.

B.3 Implications

In sum, these preliminary tests find ambiguous results at best for the influence of post-materialism and income variance on environmental quality. Part of the cause behind the poor results may be data quality. For several of the indicators, the samples were extremely small, which influenced the performance of the other variables as well. One cannot blame all of the results on poor data, however, especially not the inconsistency regarding the form of the identified relationships. Again, a more differentiated analysis of the relationship between income distributions and environmental quality might be necessary. Income distributions are likely to have a different influence on environmental quality depending on whether environmental quality tends to improve or deteriorate with increases in per capita income, for instance. If the environmental quality generally deteriorates with rising per capita incomes, as has been shown for the case of waste, for example, an income distribution that is strongly skewed towards a wealthy elite may lead to less pollution in the form of waste, as a larger share of the population will consume less and have less resources to waste. In contrast, when dealing with an indicator that generally improves with per capita income levels, a similarly skewed income distribution will lead to an overall poorer environmental quality. In the case of access to safe drinking water, for instance, a highly skewed income distribution will mean that a larger share of the population does not have this access. Similar to the debates on government regimes and centralization, then, arguments regarding the relationship between environmental quality and income variance need to be made in the context of the specific environmental dimension and indicator in question.

ENVIRONMENT & POLICY

1. Dutch Committee for Long-Term Environmental Policy: *The Environment: Towards a Sustainable Future.* 1994 ISBN 0-7923-2655-5; Pb 0-7923-2656-3
2. O. Kuik, P. Peters and N. Schrijver (eds.): *Joint Implementation to Curb Climate Change.* Legal and Economic Aspects. 1994 ISBN 0-7923-2825-6
3. C.J. Jepma (ed.): *The Feasibility of Joint Implementation.* 1995 ISBN 0-7923-3426-4
4. F.J. Dietz, H.R.J. Vollebergh and J.L. de Vries (eds.): *Environment, Incentives and the Common Market.* 1995 ISBN 0-7923-3602-X
5. J.F.Th. Schoute, P.A. Finke, F.R. Veeneklaas and H.P. Wolfert (eds.): *Scenario Studies for the Rural Environment.* 1995 ISBN 0-7923-3748-4
6. R.E. Munn, J.W.M. la Rivière and N. van Lookeren Campagne: *Policy Making in an Era of Global Environmental Change.* 1996 ISBN 0-7923-3872-3
7. F. Oosterhuis, F. Rubik and G. Scholl: *Product Policy in Europe: New Environmental Perspectives.* 1996 ISBN 0-7923-4078-7
8. J. Gupta: *The Climate Change Convention and Developing Countries: From Conflict to Consensus?* 1997 ISBN 0-7923-4577-0
9. M. Rolén, H. Sjöberg and U. Svedin (eds.): *International Governance on Environmental Issues.* 1997 ISBN 0-7923-4701-3
10. M.A. Ridley: *Lowering the Cost of Emission Reduction: Joint Implementation in the Framework Convention on Climate Change.* 1998 ISBN 0-7923-4914-8
11. G.J.I. Schrama (ed.): *Drinking Water Supply and Agricultural Pollution.* Preventive Action by the Water Supply Sector in the European Union and the United States. 1998 ISBN 0-7923-5104-5
12. P. Glasbergen: *Co-operative Environmental Governance: Public-Private Agreements as a Policy Strategy.* 1998 ISBN 0-7923-5148-7; Pb 0-7923-5149-5
13. P. Vellinga, F. Berkhout and J. Gupta (eds.): *Managing a Material World.* Perspectives in Industrial Ecology. 1998 ISBN 0-7923-5153-3; Pb 0-7923-5206-8
14. F.H.J.M. Coenen, D. Huitema and L.J. O'Toole, Jr. (eds.): *Participation and the Quality of Environmental Decision Making.* 1998 ISBN 0-7923-5264-5
15. D.M. Pugh and J.V. Tarazona (eds.): *Regulation for Chemical Safety in Europe: Analysis, Comment and Criticism.* 1998 ISBN 0-7923-5269-6
16. W. Østreng (ed.): *National Security and International Environmental Cooperation in the Arctic – the Case of the Northern Sea Route.* 1999 ISBN 0-7923-5528-8
17. S.V. Meijerink: *Conflict and Cooperation on the Scheldt River Basin.* A Case Study of Decision Making on International Scheldt Issues between 1967 and 1997. 1999 ISBN 0-7923-5650-0
18. M.A. Mohamed Salih: *Environmental Politics and Liberation in Contemporary Africa.* 1999 ISBN 0-7923-5650-0
19. C.J. Jepma and W. van der Gaast (eds.): *On the Compatibility of Flexible Instruments.* 1999 ISBN 0-7923-5728-0
20. M. Andersson: *Change and Continuity in Poland's Environmental Policy.* 1999 ISBN 0-7923-6051-6

ENVIRONMENT & POLICY

21. W. Kägi: *Economics of Climate Change: The Contribution of Forestry Projects.* 2000 ISBN 0-7923-6103-2
22. E. van der Voet, J.B. Guinée and H.A.U. de Haes (eds.): *Heavy Metals: A Problem Solved?* Methods and Models to Evaluate Policy Strategies for Heavy Metals. 2000 ISBN 0-7923-6192-X
23. G. Hønneland: *Coercive and Discursive Compliance Mechanisms in the Management of Natural Resourses.* A Case Study from the Barents Sea Fisheries. 2000 ISBN 0-7923-6243-8
24. J. van Tatenhove, B. Arts and P. Leroy (eds.): *Political Modernisation and the Environments.* The Renewal of Environmental Policy Arrangements. 2000 ISBN 0-7923-6312-4
25. G.K. Rosendal: *The Convention on Biological Diversity and Developing Countries.* 2000 ISBN 0-7923-6375-2
26. G.H. Vonkeman (ed.): *Sustainable Development of European Cities and Regions.* 2000 ISBN 0-7923-6423-6
27. J. Gupta and M. Grubb (eds.): *Climate Change and European Leadership.* A Sustainable Role for Europe? 2000 ISBN 0-7923-6466-X
28. D. Vidas (ed.): *Implementing the Environmental Protection Regime for the Antarctic.* 2000 ISBN 0-7923-6609-3; Pb 0-7923-6610-7
29. K. Eder and M. Kousis (eds.): *Environmental Politics in Southern Europe: Actors, Institutions and Discourses in a Europeanizing Society.* 2000 ISBN 0-7923-6753-7
30. R. Schwarze: *Law and Economics of International Climate Change Policy.* 2001 ISBN 0-7923-6800-2
31. M.J. Scoullos, G.H. Vonkeman, I. Thornton, and Z. Makuch: *Mercury - Cadmium-Lead: Handbook for Sustainable Heavy Metals Policy and Regulation.* 2001 ISBN 1-4020-0224-6
32. G. Sundqvist: *The Bedrock of Opinion.* Science, Technology and Society in the Siting of High-Level Nuclear Waste. 2002 ISBN 1-4020-0477-X
33. P.P.J. Driessen and P. Glasbergen (eds.): *Greening Society.* The Paradigm Shift in Dutch Environmental Politics. 2002 ISBN 1-4020-0652-7
34. D. Huitema: *Hazardous Decisions.* Hazardous Waste Siting in the UK, The Netherlands and Canada. Institutions and Discourses. 2002 ISBN 1-4020-0969-0
35. D. A. Fuchs: *An Institutional Basis for Environmental Stewardship: The Structure and Quality of Property Rights.* 2003 ISBN 1-4020-1002-8

For further information about the series and how to order, please visit our Website
http://www.wkap.nl/series.htm/ENPO

KLUWER ACADEMIC PUBLISHERS – DORDRECHT / BOSTON / LONDON

MIX
Papier aus verantwortungsvollen Quellen
Paper from responsible sources
FSC® C105338

If you have any concerns about our products,
you can contact us on
ProductSafety@springernature.com

In case Publisher is established outside the EU,
the EU authorized representative is:
Springer Nature Customer Service Center GmbH
Europaplatz 3, 69115 Heidelberg, Germany

Printed by Libri Plureos GmbH
in Hamburg, Germany